FRONTIERS IN WATER RESOURCE ECONOMICS

NATURAL RESOURCE MANAGEMENT AND POLICY

Editors:
Ariel Dinar
Rural Development Department
The World Bank
1818 H Street, NW
Washington, DC 20433

David Zilberman
Dept. of Agricultural and
Resource Economics
Univ. of California, Berkeley
Berkeley, CA 94720

EDITORIAL STATEMENT

There is a growing awareness to the role that natural resources such as water, land, forests and environmental amenities play in our lives. There are many competing uses for natural resources, and society is challenged to manage them for improving social well being. Furthermore, there may be dire consequences to natural resources mismanagement. Renewable resources such as water, land and the environment are linked, and decisions made with regard to one may affect the others. Policy and management of natural resources now require interdisciplinary approach including natural and social sciences to correctly address our society preferences.

This series provides a collection of works containing most recent findings on economics, management and policy of renewable biological resources such as water, land, crop protection, sustainable agriculture, technology, and environmental health. It incorporates modern thinking and techniques of economics and management. Books in this series will incorporate knowledge and models of natural phenomena with economics and managerial decision frameworks to assess alternative options for managing natural resources and environment.

The Series Editors

Recently Published Books in the Series

Haddadin, Munther J.
Diplomacy on the Jordan: International Conflict and Negotiated Resolution
Renzetti, Steven
The Economics of Water Demands
Just, Richard E. and Pope, Rulon D.
A Comprehensive Assessment of the Role of Risk in U.S. Agriculture
Dinar, Ariel and Zilberman, David
Economics of Water Resources: The Contributions of Dan Yaron
Ünver,.I.H. Olcay, Gupta, Rajiv K. IAS, and Kibaroğlu, Ayşegül
Water Development and Poverty Reduction
d'Estrée, Tamra Pearson and Colby, Bonnie G.
Braving the Currents: Evaluating Environmental Conflict Resolution in the River Basins of the American West
Cooper, Joseph, Lipper, Leslie Marie and Zilberman, David
Agricultural Biodiversity and Biotechnology in Economic Development

FRONTIERS IN WATER RESOURCE ECONOMICS

Edited by

RENAN-ULRICH GOETZ
University of Girona

DOLORS BERGA
University of Girona

Springer

Library of Congress Control Number: 2005935082

ISBN:10: 0-387-30055-4
ISBN-13: 978-0387-30055-9
e-ISBN-10: 0-387-30056-2
e-ISBN-13: 978-0387-30056-6

Printed on acid-free paper.

Printed in the United States of America.

9 8 7 6 5 4 3 2 1

springeronline.com

Contents

List of Figures

List of Tables

Contributing Authors

Stephen Beare, Chief Economist, Australian Bureau of Agricultural and Resource Economics, Canberra, Australia

Dolors Berga, Associate Professor, Departament d'Economia, Universitat de Girona, Girona, Spain

Nicholas Brozović, Assistant Professor, Department of Agricultural and Consumer Economics, University of Illinois at Urbana-Champaign, Urbana, Illinois, U.S.A.

Javier Calatrava Leyva, Associate Professor, Departamento de Economía de la Empresa, Universidad Politécnica de Cartagena, Cartagena, Spain

Ephraïm Clark, Professor of Finance, Department of Finance, Middlesex University, London, UK

Daniel Cohen-Vogel, Senior Policy Analyst, Florida Legislature, Office of Program Policy Analysis and Government Accountability, Tallahassee, Florida, U.S.A.

Ariel Dinar, Lead Economist, Department of Agriculture and Rural Development, World Bank, Washington DC, U.S.A.

Richard Farnsworth, Associate Professor, Department of Forestry and Natural Resources, Purdue University, West Lafayette, Indiana, U.S.A.

Linda Fernandez, Associate Professor, Department of Environmental Sciences, University of California at Riverside, Riverside, California, U.S.A.

Alberto Garrido, Associate Professor, Department of Agricultural and Resource Economics, Universidad Politécnica de Madrid, Madrid, Spain

Renan-Ulrich Goetz, Associate Professor, Departament of Economics, Universitat de Girona, Girona, Spain

Anna Heaney, Manager, Department of Land & Water Economics; Natural Resources Branch, Australian Bureau of Agricultural and Resource Economics, Canberra, Australia

Petra J.G.J. Hellegers, Senior Researcher, Department of Management of Natural Resources, Agricultural Economics Research Institute (LEI) (Wageningen University and Research Institute), Wageningen, The Netherlands

Madhu Khanna, Professor, Department of Agricultural and Consumer Economics, University of Illinois at Urbana-Champaign, Urbana, Illinois, U.S.A.

Jean-Jacques Laffont†**,** Professor, Université de Toulouse, Toulouse, and Centre National de la Recherche Scientifique, France; and University of Southern California, U.S.A.

Gérard Mondello, Researcher, Groupe de Recherche en Droit, Economie, Gestion; Centre National de la Recherche Scientifique, Sophia Antipolis, France

Georgina A. Moreno, Assistant Professor, Department of Economics, Scripps College, Claremont, California, U.S.A.

Stefano Moretti, Ph.D. student, Department of Mathematics, University of Genoa and Unit of Molecular Epidemiology, National Cancer Research Institute of Genoa, Genoa, Italy

Fioravante Patrone, Full Professor, Dipartimento di Ingegneria della Produzione, Termoenergetica e Modelli Matematici, University of Genoa, Genoa, Italy

Jeanne Reeves, Director, Agricultural Research, Cotton Incorporated, Cary, North Carolina, U.S.A.

Catarina Roseta-Palma, Assistant Professor, Dinâmia e Departâmento de Economia, Instituto Superior de Ciências do Trabalho e da Empresa, Lisbon, Portugal

François Salanié, Chargé de Recherche, Institut National de la Recherche Agronomique, Laboratoire d'Economie des Ressources Naturelles, Université de Toulouse, Toulouse, France.

David Sunding, Professor, Department of Agricultural and Resource Economics, University of California at Berkeley, Berkeley, California, U.S.A.

Àngels Xabadia, Assistant Professor, Departamento de Análisis Económico, Universidad de Zaragoza, Zaragoza, Spain

Stefano Zara, Collaborator, University of Genova, Iglesias (Cagliari), Italy

David Zilberman, Professor and Robinson Chair, Department of Agricultural and Resource Economics, Director, Giannini Foundation of Agricultural Economics, University of California at Berkeley, Berkeley, California, U.S.A.

Acknowledgments

The editors of this book would like to thank the company CESPA Ingeniería Urbana and the Institute of the Environment of the University of Girona for their financial support of the congress (7th Conference of the International Water and Resource Economics Consortium and 4th Seminar on Environmental and Resource Economics) that gave rise to a large number of papers collected in this book. We would also like to extend our thanks to Àngels Xabadia who unified the layout of the different chapters in a timely and excellent manner. Finally our thanks go to the anonymous reviewers for their comments and suggestions that helped to improve the readability of the text and the presentation of complex issues.

Introduction

Most of the books published previously in the field of water resource economics focus on particular aspects of water economics such as institutions, pricing or water markets, but none of them have given particular attention to methodological questions. However, the applied methodology within economic research has made some remarkable advances over the last 10 - 20 years. Some of these advances are of particular interest to the field of water economics. Therefore, we think that a book that focusing on methodological advances within the field of water resource economics and showing how these advances can be applied in economic analysis of water issues makes a nice complement to the existing literature in this field.

We identified five areas where we consider the methodological advances to be of particular importance: 1) asymmetric information and game theory, 2) uncertainty, 3) space, 4) water quality and 5) production and technology adoption. The selected papers for the book fall entirely within these categories. The book *"Frontiers in Water Resource Economics"* draws to a great extent on papers which were presented at the *7th Conference of the International Water and Resource Economics Consortium*, June 3-5, 2001 held in Girona, Catalonia, Spain. This conference was jointly organized with *the 4th Conference of Environmental and Resource Economics* by the Department of Economics, University of Girona. The chapters in this book have been selected, in part, from the above mentioned conferences and solicited, in part, from distinguished authors in the field. The composition of the book is based on the five areas identified above. In what follows we briefly present the different chapters for each area.

Asymmetric information and game theory (Chapters 1 - 2)

The paper by Dinar, Moretti, Patrone, and Zara, "Application of Stochastic Cooperative Game Theory in Water Resources", applies concepts of stochastic cooperative games to water resource problems. Water resource projects are subject to stochastic water supply patterns, which affect their performance, sustainability and the stability of any use arrangement among participants. Traditionally, cooperative game theory has been applied to a variety of water resource problems assuming a deterministic pattern of supply. By means of different examples the authors show how incorporation of stochastic considerations may change the solution, depending on the attitude toward risk aversion of the potential users and on the nature of the cost function of the joint water project. Another chapter within this area is by Laffont and Salanié, titled "Incentives and the Search for Unknown Resources such as Water". This work uses the principal-agent methodology to analyze the problems in contracting agents to invest in the search for a natural resource such as water, when these agents are informed about the size of their discoveries but the other parties are not. The chapter is related to three strands of the literature: the principal-agent models with type-dependent status quo payoffs, law and economics literature and the mechanism design literature about the manipulation of endowments. A remarkable result is that limited liability constraints and enforcement problems result in unusual distortions compared to the first best as the principal tries to decrease the agent's effort.

Uncertainty (Chapters 3 - 5):

The chapter by Calatrava Leyva and Garrido, "Risk Aversion and Gains from Water Trading under Uncertain Water Availability", analyzes on theoretical and empirical grounds the impact of risk-averse behavior and uncertainty of water supply on the efficiency gains from spot water markets. It examines the role that uncertainty and risk aversion play in market participants' decisions and their impact on gains from trade. Irrigators' behavior is simulated using a two-stage discrete stochastic programming model, while market exchanges and equilibrium are computed using a price endogenous spatial equilibrium model. Participation in the water market is modeled as a tactical response, and a utility-efficient modeling approach is used to account for risk aversion. The empirical application is performed simulating a local water market in an irrigation district of the Guadalquivir Valley (Southern Spain). Results show that gains from trade diminish as uncertainty and risk-aversion levels increase.

The chapter by Clark and Mondello, "Dynamic Uncertainty and the Pricing of Natural Monopolies: The Case of Urban Water Management", presents a model based on techniques utilized in real option theory. The model can be used as a pricing program that gives a fair deal to both the regulator and the delegated firm. Their pricing program goes beyond marginal cost and non-linear pricing by introducing time and risk. Importantly, it provides a solution to the price that the delegated firm should pay for the right to exploit the monopoly. It also includes the relatively recent problem of technological monopolies that make it possible for the private water firms to extract supplementary economic rents. The fair deal in their pricing program eliminates economic rents while rewarding the risk taken on by the firm. It also guarantees that the regulator receives the full value of the monopoly that it is ceding to the firm. In this sense the fair deal that they propose is an optimum.

The final chapter in this area by Moreno and Sunding, "Price Risk and the Diffusion of Conservation Technology", explores the diffusion pattern of conservation technology in irrigated agriculture. The chapter examines the influence of factor price risk on factor-use efficiency through the adoption of conservation technology. Conceptual results indicate that a mean-preserving increase in factor price risk has an ambiguous effect on returns to investment in conservation technology, but should be related to the own-price elasticity of input utilization. Theoretical results are tested by estimating an ordered probit model of technology choice using a unique data set from the water industry. Estimation results are consistent with the conceptual model, and also indicate that price risk can have a large influence on the adoption decision.

Space (Chapters 6 -8):

The chapter by Brozović, Sunding, and Zilberman, "Optimal Management of Groundwater over Space and Time", presents a model for the optimal management over space and time of a groundwater resource with multiple users. Their model extends the existing literature, incorporating both space and time and using the hydraulic response equations that govern the behavior of groundwater. In their analysis, the authors emphasize how physical parameters of the groundwater system affect the spatial and temporal distribution of extraction. A discussion of the optimality conditions from their model emphasizes how the results differ from existing studies and the policy implications of these differences. In particular, they show that some aquifers are more akin to private property than common property, and that there may be significant lagged effects from pumping.

The chapter by Goetz, Berga, and Xabadia, "Nonpoint Source Pollution in a Spatial Intertemporal Context - a Deposit Refund Approach", shows that the incorporation of space allows a more precise relationship to be established between a contaminating byproduct and the emissions that reach the final receptor. However, the presence of asymmetric information impedes the implementation of the first-best policy. As a partial solution to their problem they propose a site specific deposit refund for the contaminating byproduct. Moreover, the utilization of a successive optimization technique, first over space and second over time, enables them to define the optimal intertemporal site specific deposit refund system.

The final chapter in this area is by Fernandez, "Transboundary Water Management along the U.S.-Mexico Border". It presents an extensive survey of the studies concerning the problem of surface and groundwater resources shared by different nations. Special attention is given to the U.S.-Mexico border, where transboundary issues vary in terms of directions of flow and policy at national and binational levels. Typically, game theory concepts help examine incentives between countries sharing water resources. There have been developments in policy structure from the historical to current context in which to assess formal channels of financing current and future water infrastructure along the border. The chapter also presents additional water analyses beyond the U.S.-Mexico border.

Water quality (Chapters 9 - 11):

The chapter by Beare and Heaney, "Irrigation, Water Quality and Water Rights in the Murray Darling Basin, Australia", focuses on the development of water markets and the establishment of property rights to facilitate trade and create efficient incentives for investing in water use efficiency. However, irrigation can impose significant costs on both downstream water users and the environment. The extent to which trade and investments in improved water use technologies will lead to efficiency gains depends, in part, on how well water property rights account for the externalities associated with irrigation. The chapter presents an economic model of irrigation agriculture that is integrated with a hydrological model of the Murray-Darling River system to determine the potential magnitude of benefits of establishing property rights which take into account the downstream impacts of return flows. Establishing site specific conditions on property rights that partially internalize the external effects of irrigation may lead to an improvement in economic welfare. These could include taxes or subsidies on both trade between irrigation regions and investment in water saving technologies in specific irrigation regions.

The next chapter by Khanna and Farnsworth, "Economic Analysis of Green Payments to Protect Water Quality", draws our attention to the economic literature on green payment policies. It provides an extensive review of both the theoretical and empirical literature on three basic issues in this field. What criteria should be used by a policy maker to select the recipients of these green payments? In a decentralized decision making situation, where participation in a green payment program is a voluntary decision of the farmer, how should these payments be designed to maximize social welfare or to achieve given environmental objectives cost-effectively? How efficient are green payment programs relative to first best instruments to control pollution that would have maximized social welfare? It shows the importance of incorporating spatial heterogeneity in the costs and benefits of alternative land uses while designing and targeting green payment policies.

The final chapter in this area is by Roseta-Palma, "Conjunctive Use of Surface Water and Groundwater with Quality Considerations". It starts from the observation that deterministic models of conjunctive surface and groundwater management are not much more complicated than typical groundwater-only management models under simple assumptions. However, when water quality problems exist, the fact that there are two alternative sources of water gains a special significance, as there is no guarantee that they will be of comparable quality. Considering that water quality varies according to source, the paper analyzes some new implications that arise in a conjunctive system, with and without uncertainty in hydrological parameters.

Production and technology adoption (Chapters 12 - 13):

The chapter by Hellegers, "The Impact of Recovering Irrigation Water Losses on the Choice of Irrigation Technology with Heterogeneous Land Quality and Different Crops", starts from the affirmation that an increase in the irrigation effectiveness - by adoption of modern irrigation technologies - is often proposed as a solution to water scarcity. Whether this is desirable in all cases is studied in this chapter. An existing conceptual farm model is used to show the impact of heterogeneous land quality on the choice of irrigation technology. Introduced in the model are recovery of irrigation water losses as well as the possibility of growing different crops on land with different quality, both extensions of existing work in this field. It becomes clear that recovery reduces resource costs -which makes adoption of modern irrigation technology less likely-, but may impose costs when recharge flows are of a bad quality -which makes adoption more likely. These costs are, however, often not paid by the farmer, which explains why adoption is sometimes subsidized.

The final chapter of the book is by Zilberman, Cohen-Vogel and Reeves. "Precision Farming in Cotton" focuses on precision farming as a new category of technology that adjusts the application level of agricultural input and that accommodates variations within fields and also climatic and other seasonal variations. The authors develop a model of the impact of these technologies, recognizing heterogeneity in terms of utility and water holding capacity within fields. This is the first quantitative study of the adoption of varying input technology. Unlike the literature on adoption of irrigation technology, they recognize the importance of within-field heterogeneity and the capacity to adjust to heterogeneity. They develop general conditions under which precision technology will be adopted and show that introduction of precision technologies may lead to water saving and yield increasing effects in some cases while in others it may lead to water saving and yield reduction. They further develop a relationship that suggests that the impacts of the technology depend on the various dimensions of heterogeneity-within, using an example from California cotton fields and assuming uniform distribution of fertility and water holding capacity. They show that under plausible conditions technology adoption may result in a 10%-40% increase in profit.

Chapter 1

APPLICATION OF STOCHASTIC COOPERATIVE GAMES IN WATER RESOURCES*

Ariel Dinar
World Bank, U.S.A.

Stefano Moretti
University of Genoa, Italy

Fioravante Patrone
University of Genoa, Italy

Stefano Zara
Via Roma, 30 - 09016 Iglesias (CA) - Italy

1. INTRODUCTION

In March 2001, the Cahora Bassa (also known as Cabora Bassa) Dam on the Zambezi River in Mozambique could not hold the huge volume of water that filled its reservoir. As a result, devastating floods wiped out thousands of communities and their infrastructure, killing hundreds and leaving a quarter of a million others homeless.

Zambia and Mozambique are riparians to the Zambezi, and the Cahora Bassa reservoir is filled with water flowing from Zambia. Dam operating rules that were set, taking into account average flow and hydropower preferences (over flood protection) are blamed for the disaster. The dam operating rule kept the

*The authors thank an anonymous referee for very detailed and constructive comments.

water spillway at 80% of the average river flow, not leaving sufficient margin for extreme quantities. Was the damage (in both lives and economic loss) prevented, had the designers take into consideration extreme and stochastic events such that the 2001 flood?

Situations similar to the Mozambique flood are very common around the world (e.g., the Elbe flooding of vast regions in the Czech Republic and Germany in Fall 2002). The problem of determining the operational level of a joint facility holds not only for dams. It is relevant also in setting the size and water quality in treatment facilities of sewerage and desalinization, and in any investment that could be affected by stochastic events. Thus, the issue of optimal design of a joint facility is of great economic importance.

Game theory (GT) contributed to the understanding of allocation of costs and benefits of joint facilities among different users. In the water sector in particular, cooperative GT work (e.g., Gately, 1974; Loheman et al., 1979; Suzuki and Nakayama, 1976; Straffin and Heany, 1981; Dinar et al., 1986; Rogers, 1993; Moretti et al., 2002) addressed various aspects of water sector issues, including hydropower, water storage, multi-objective water projects, municipal sewage treatment, disposal and reuse, and international water cooperation. Most of the work, applying cooperative GT depart from the assumption of a deterministic world. As such, the proposed GT solutions to the allocation problems are probably restricted to a very narrow subset of average behavior of the events that affect the main input to the allocation problem, water.

Dinar and Howitt (1997) attempted, in a very naive way, to address the problem of stochastic supply of water, affecting the design of a joint facility of drainage water treatment in the San Joaquin Valley in California. Although very specific to their problem, Dinar and Howitt identified that the stability of the allocation arrangement is sensitive to the state of nature, depending also on the selected allocation scheme.

In view of the important role that water plays in regional and local projects, and taking into account that with climate change affecting the water cycle, the world is expected to face more stochastic and extreme events of water supply, incorporating stochastic consideration of water supply becomes more acute in designing water facilities. We shall provide a couple of examples (see Section 3) to show the effects of variability in water supply on the cooperative game used to model the cost allocation problem. Moreover, various water users may have different attitudes toward risk, depending on their economic, managerial, and institutional capacity. Therefore, the combination of stochastic events and players' risk attitude becomes increasingly an important issue in designing water related economic activities that depend on cooperation among the users.

In this chapter we will apply a Stochastic GT framework, based on the work of Suijs and Borm (1999). Stochastic cooperative games have been applied in some fields (insurance: Suijs et al., 1998; network enterprises: Suijs, 2003; risk trade for recovery projects after a disaster: Tanimoto et al., 2000). We consider that this analytical tool could be of use also in water-related problems, when variability in the input is high. We will use an example of a water treatment plant to illustrate the approach, although the principles hold also for cases of other water-related joint cost, such as for storage, etc.

Our goal will be to offer, in a simplified setting, some examples of the problems which appear when one tries to take into account the stochastic aspects of the problem which is being modelled. We shall also show, by means of an example, that risk aversion of the players may influence the possibility of getting allocations which lie in the core of the game which models the cost allocation problem.

Section 2 will introduce the general model, mainly describing the cost function that will be used. Section 3 describes cooperative games which arise under a couple of different interpretations of the cost function. Section 4 analyzes the core of a cooperative stochastic game, showing the relevance of risk aversion of the players. Section 5 concludes.

2. THE GENERAL MODEL

Assume a finite set of possible states of nature: $\Omega = \{\omega_1, \ldots, \omega_k\}$, associated with a probability distribution $p(\omega_i)$.

Assume also a cost function

$$C : [0, +\infty[\to [0, +\infty[,$$

about which we shall offer a couple of different interpretations.

Assume further a given set of players (in this chapter we will use interchangeably the terms players and users) $N = \{1, \ldots, n\}$.

Each one of these players has to deal with some amount of polluted water, whose quantity depends also on the state of nature. For a given quantity of polluted water, the function C describes the cost for its treatment.

In general, C will be derived by the application of an appropriate technology to the amount of water to be treated, and will depend also on a vector G of environmental variables such as water quality, soil properties, landscape, etc. For a given G, we assume that $C = C(q)$ is obtained by choosing the most appropriate (less costly) technology.

Notice that C can be affected by selecting the amount q of water to be applied and the technology in response to different states of nature (see for example Dinar and Zilberman, 1991; Dinar et al., 1992). Our general model does not allow for such responses at this stage. The main reason is that our model aims to address mainly ex-ante situations in various sectors. It can represent irrigation and drainage issues, municipal water supply and sewerage, hydropower and flood control, and industrial water use and sewerage. The model deals with ex-ante planning and management of the joint facility. For these two reasons it is assumed that both q and the technology t have to take into account the state of nature in the planning stage and cannot not be adjusted to it instantaneously. On the other hand, a different interpretation of the set of available technologies as the set of possible actions for the players of a stochastic cooperative game is provided in Section 4 (see Example).

We assume also a function which describes for every player the quantity of polluted water that he has to deal with, for any given state of nature.

A simplified version assumes that $q(i,\omega) = \phi(i)\hat{q}(\omega)$. This will be the case that we shall discuss in the Example. An almost obvious interpretation for this case is to assume that the quantity $q(i,\omega)$ of polluted water that each player has to deal with equals the product between the amount of water for unit area $\hat{q}(\omega)$ that reaches the soil (depending, trivially, on the weather conditions) and an idiosyncratic factor $\phi(i)$ representing the water to be treated (depending, for instance, on the amount, quality and characteristics of the soil and on the kind of cultivations).

We are interested in showing how the players can jointly reduce the costs of treatment, and find reasonable ways in which they can divide the resulting joint costs; this procedure will be accomplished through the definition of an appropriate transfer utility game (TU-game).

In a given state of nature $\omega \in \Omega$, the treatment cost for a given set of players S is given by $C\left(\sum_{i\in S} q(i,\omega)\right)$.

Moving to the aggregation on the states of nature, we suggest here, and develop in Section 3 (see considerations *a)* and *b)*), two different interpretations which can be of interest, each one of which leads to a TU-game.

One interpretation is that

$$\hat{c}(S) = \max_{\omega\in\Omega} C\left(\sum_{i\in S} q(i,\omega)\right).$$

Another is that

$$\bar{c}(S) = \sum_{\omega\in\Omega} p(\omega)C\left(\sum_{i\in S} q(i,\omega)\right),$$

where $p(\omega)$, $\omega \in \Omega$, is a probability distribution on the set of states of nature Ω.

In the case of a stochastic TU-game, a relevant factor is risk aversion of the players (more precisely, the different degrees of risk aversion that players may exhibit). In this chapter we focus on the case of risk neutral players; we provide just an illustrative example of a stochastic TU-game in which risk aversion of players, relaxing the stability conditions, can provide a non empty core, while risk indifference of the players would imply that the core is empty.

In the case of risk neutral players, we use step cost functions as a basis for the cost calculations. These functions, which we study in Section 3, are not concave, so that it will be no surprise that we can offer examples in which the core is empty or that the Shapley value lies outside the core. Step cost functions are clearly a simplified approach, but we believe that they contain one essential feature of the problem, with regard to stochastic factors: it can be assumed that the cost of water treatment is a concave function of the quantity of water use, but this realistically holds only for some range of the quantity. Whenever there is a strong variance in the water use input, it is difficult to land in a unique system of water treatment, which would give the concavity of the cost function. It is sound to use a given setup to treat water, but with some limits on the capacity, while for greater amounts of water to be treated it could be more appropriate to use other setups. Such setups could be more expensive in layout costs involved, or because they use a different technology. The alternative to using a setup that gives an answer to any quantity of polluted water, would mean to have an oversized treatment plant for almost all of the states of nature, and thus a much more expensive one, due to the inherent high fixed costs.

In the following sections we detail these situations, providing numerical examples of the phenomena that we have described (empty core, Shapley value not in the core[1]).

3. WATER GAMES

Water users have to tackle the ex-ante decision problem of how to share costs of water treatment (costs which are univocally realized only ex-post). What they precisely know is the probability distribution on occurrence of the states of nature and the cost function of treating water, which is depicted in Figure 1. We want to state clearly that we consider the simplest case of cost function that still exhibits interesting behavior from the viewpoint of cost allocation. It would be very easy to extend the analysis to more general cost functions (including more steps) than the one described in Figure 1

We argue that the ex-ante agreement between the players on the cost sharing rule to be adopted ex-post, is based just on the cost function and the probability distribution, and maybe also on the risk attitude of the players. So, it is very important to understand what is the meaning of a cost function such as the one depicted in Figure 1. We try to offer two different interpretations of the function in Figure 1 with regard to the case under consideration:

a) A plant for water treatment has to be built. Solving this decision problem provides the cost function for the users. We assume that this decision problem will be solved according to the following criterion: if the amount of water to be treated is smaller than Q^*, then a small plant should be built and the cost supplied is C_1; otherwise if the amount of water to be treated is larger than or equal to the amount Q^*, then a bigger plant must

be built and the cost is C_2 ($C_1 < C_2$).
Moreover, in this case we assume that for a given group of players $S \subseteq N$, if a state of nature $\bar{\omega}$ s.t. $\sum_{i \in S} q(i, \bar{\omega}) \geq Q^*$ is possible (i.e. the probability of $\bar{\omega}$ to occur is larger than 0), then the group will decide to construct the larger plant and their cost is C_2 (in fact, avoiding to build the bigger plant could be very dangerous, since users could face a situation of over-pollution). Otherwise, if for each state $\omega \in \Omega$ with probability different than 0 the corresponding amount of polluted water $q(\omega)$ is such that $\sum_{i \in S} q(i, \omega) < Q^*$, then the group will build the smaller plant and the cost will be C_1;

b) A plant for water treatment has already been built. Then for each state of nature $\omega \in \Omega$ and for any group of players S such that $\sum_{i \in S} q(i, \omega) < Q^*$, the cost of water treatment in the given plant is C_1. If a group of players S has to treat a quantity $\sum_{i \in S} q(i, \omega) \geq Q^*$ in some states of nature $\omega \in \Omega$, the cost becomes C_2 in such states of nature. In this interpretation, the increase in cost could be due to some facilities/services needed only when the amount Q^* has been exceeded (for example the renting of tankers used to transport liquid to another place).

Note that these are "extreme" cases that will be modelled using the treatment cost functions defined in Section 2, $\hat{c}$ and $\bar{c}$ respectively. Our analysis could be extended to other situations in the sense that we could be able to provide

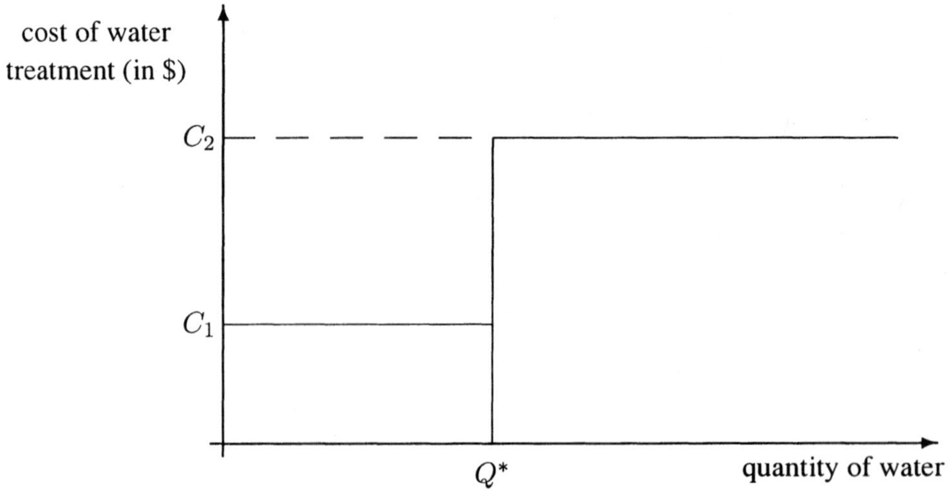

Figure 1. Cost function of water treatment

examples analogous to those we have just introduced to treat more general functions as in Figure 2. In this figure it is shown a case in which the cost function continuously increases with the quantity of water that is affected by different states of nature (e.g. due to an increment of rainfall events which implies a consequent growth in terms of facilities/services for water treatment) and again there is a quantity Q^* that, if reached, requires the ex-ante construction of a new plant (with corresponding cost increases).

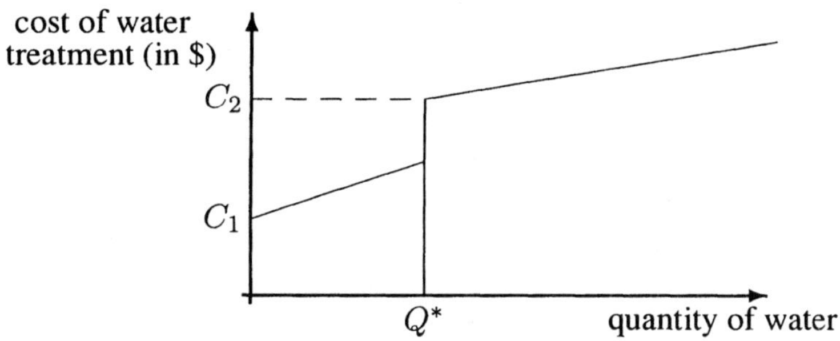

Figure 2. An intermediate situation between interpretation (a) and (b)

In the following sections, we will refer to a cost function like the one depicted in Figure 1, in order to define two cooperative games, following the interpretations (a) and (b), respectively.

3.1 Considerations on interpretation (a)

Consider the case with 3 players $N = \{1, 2, 3\}$ and a set of 3 states of nature $\Omega = \{\omega_1, \omega_2, \omega_3\}$.

Suppose that $q(i, \omega) = \phi(i)\hat{q}(\omega)$, where $\phi(1) = \phi(2) = 1$, $\phi(3) = 2$ (reflecting three farms with the land for player 3 being twice as big as for players 1 and 2), and $\hat{q}(\omega_i) = Q_i$, describes the amount of water to be treated. For each state $\omega \in \Omega$ and for each player 1, 2, 3, values are reported in the following table:

state of the world \ players	ω_1	ω_2	ω_3
1	Q_1	Q_2	Q_3
2	Q_1	Q_2	Q_3
3	$2Q_1$	$2Q_2$	$2Q_3$

We furthermore assume that $Q_1 = 2$, $Q_2 = 3$, $Q_3 = 4$ and $Q^* = 10.9$ (here Q^* refers to Figure 1). Moreover we suppose that there is a given probability distribution on the states $\omega_1, \omega_2, \omega_3$ different from zero for each sate. Notice that in this interpretation the quantification of the probability attached to the states of nature is not relevant, as long as it can be assumed positive for each of them. Then, by interpretation (a), the characteristic cost function values for each state of nature for the resulting cooperative game of sharing the joint water treatment cost are summarized in the following schema:

	ω_1	ω_2	ω_3	max cost
$\{1\}$	C_1	C_1	C_1	C_1
$\{2\}$	C_1	C_1	C_1	C_1
$\{3\}$	C_1	C_1	C_1	C_1
$\{1,2\}$	C_1	C_1	C_1	C_1
$\{1,3\}$	C_1	C_1	C_1	C_1
$\{2,3\}$	C_1	C_1	C_2	C_2
$\{1,2,3\}$	C_1	C_2	C_2	C_2

For example, consider coalition $\{2,3\}$: the cost for the plant under state ω_2 is C_1 since $q(2,\omega_2) + q(3,\omega_2) = Q_2 + 2Q_2 = 9 < Q^*$, whereas under state ω_3 the cost will be C_2 because $q(2,\omega_3) + q(3,\omega_3) = Q_3 + 2Q_3 = 12 > Q^*$. Therefore, coalition $\{2,3\}$ could face an amount of polluted water greater than Q^* (i.e. state ω_3). Therefore, the cost for coalition $\{2,3\}$ in the resulting (ex-ante) game should be the maximum cost C_2.

We are going to consider the resulting (ex-ante) game $\langle\{1,2,3\},\hat{c}\rangle$, where $\hat{c}(\emptyset) = 0$, $\hat{c}(\{1\}) = C_1$, $\hat{c}(\{2\}) = C_1$, $\hat{c}(\{3\}) = C_1$, $\hat{c}(\{1,2\}) = C_1$, $\hat{c}(\{1,3\}) = C_1$, $\hat{c}(\{2,3\}) = C_2$ and $\hat{c}(\{1,2,3\}) = C_2$.

This game is not concave. In fact a cost game $\langle N, c\rangle$ is concave if for each $i \in \{1,2,3\}$ and each S, T such that $S \subseteq T \subseteq N \setminus \{i\}$

$$c(T \cup \{i\}) - c(T) \leq c(S \cup \{i\}) - c(S), \tag{1}$$

which is not satisfied for coalitions $S = \{1\}$ and $T = \{1,2\}$ and $i = 3$.
The core $Core(\{1,2,3\},\hat{c})$ of the game $\langle\{1,2,3\},\hat{c}\rangle$ is given by all the vectors $(x_1, x_2, x_3) \in \mathbb{R}^3$ such that:

$$\begin{cases} x_i \leq C_1 & \forall i \in \{1,2,3\} \\ x_1 + x_2 \leq C_1 \\ x_1 + x_3 \leq C_1 \\ x_2 + x_3 \leq C_2 \\ x_1 + x_2 + x_3 = C_2 \end{cases}$$

Then, we can write:

$$\begin{cases} x_i \leq C_1 & \forall i \in \{1,2,3\} \\ x_1 + x_2 + x_3 - x_3 \leq C_1 \\ x_1 + x_3 + x_2 - x_2 \leq C_1 \\ x_2 + x_3 + x_1 - x_1 \leq C_2 \\ x_1 + x_2 + x_3 = C_2 \end{cases}$$

$$\begin{cases} x_i \leq C_1 & \forall i \in \{1,2,3\} \\ C_2 - x_3 \leq C_1 \\ C_2 - x_2 \leq C_1 \\ C_2 - x_1 \leq C_2 \\ x_1 + x_2 + x_3 = C_2 \end{cases}$$

$$\begin{cases} C_1 \geq x_3 \geq C_2 - C_1 \\ C_1 \geq x_2 \geq C_2 - C_1 \\ C_1 \geq x_1 \geq 0 \\ x_1 + x_2 + x_3 = C_2 \end{cases}$$

Finally
$C_2 = x_1 + x_2 + x_3 \geq 2(C_2 - C_1) + 0 \Leftrightarrow C_2 \geq 2C_2 - 2C_1 \Leftrightarrow 2C_1 \geq C_2$.
So if $C_2 > 2C_1$ then *Core*$(\{1,2,3\}, \hat{c})$ is empty.

Consider now the Shapley value. Refer to the following table concerning the marginal contribution of players $1, 2, 3$ for each permutation of coalition formation:

	1	2	3
123	C_1	0	$C_2 - C_1$
132	C_1	$C_2 - C_1$	0
213	0	C_1	$C_2 - C_1$
231	0	C_1	$C_2 - C_1$
312	0	$C_2 - C_1$	C_1
321	0	$C_2 - C_1$	C_1
total	$2C_1$	$3(C_2 - C_1) + 2C_1$	$3(C_2 - C_1) + 2C_1$
$\phi_i =$	$\frac{C_1}{3}$	$\frac{C_1}{3} + \frac{C_2 - C_1}{2}$	$\frac{C_1}{3} + \frac{C_2 - C_1}{2}$

Is Shapley value in *Core*$(\{1,2,3\}, \hat{c})$? For individual rationality of players $1, 2, 3$:
$\phi_2 = \phi_3 = \frac{C_1}{3} + \frac{C_2 - C_1}{2} \leq C_1 \Leftrightarrow 2C_1 + 3C_2 - 3C_1 \leq 6C_1 \Leftrightarrow 3C_2 \leq 7C_1$
$\phi_1 = \frac{C_1}{3} < C_1$, individual rationality for player 1 is always satisfied.
For coalitional rationality of coalitions $\{1,3\}$ and $\{1,2\}$:
$\phi_1 + \phi_3 = \phi_1 + \phi_2 = \frac{2C_1}{3} + \frac{C_2 - C_1}{2} \leq C_1 \Leftrightarrow 4C_1 + 3C_2 - 3C_1 \leq 6C_1 \Leftrightarrow$
$3C_2 \leq 5C_1$
and of coalition $\{2,3\}$

$\phi_2 + \phi_3 = \frac{2C_1}{3} + C_2 - C_1 \leq C_2$, which is always satisfied.
So, if $C_2 > \frac{5}{3}C_1$ then the Shapley value is not in $Core(\{1,2,3\},\hat{c})$.

3.1.1 On Bondareva-Shapley and on the Shapley value not in the core

Remember that we are considering the resulting (ex-ante) game: $\langle\{1,2,3\},\hat{c}\rangle$, where $\hat{c}(\emptyset) = 0$, $\hat{c}(\{1\}) = C_1$, $\hat{c}(\{2\}) = C_1$, $\hat{c}(\{3\}) = C_1$, $\hat{c}(\{1,2\}) = C_1$, $\hat{c}(\{1,3\}) = C_1$, $\hat{c}(\{2,3\}) = C_2$, $\hat{c}(\{1,2,3\}) = C_2$.

To check whether the core is not empty, we can apply the Bondareva Shapley theorem (cf. Bondareva, 1963 and Shapley, 1967). The minimal balanced sets of coalition for a 3 person game are given by the possible partitions of $\{1,2,3\}$ plus the collection $\{\{1,2\},\{1,3\},\{2,3\}\}$, with balancing weights 1/2. Applying the balancedness conditions, we get:

$$\begin{cases} C_1 + C_1 + C_1 \geq C_2 & \text{for the collection } \{\{1\},\{2\},\{3\}\} \\ C_1 + C_2 \geq C_2 & \text{for the collection } \{\{1\},\{2,3\}\} \\ C_1 + C_1 \geq C_2 & \text{for the collection } \{\{2\},\{1,3\}\} \\ C_1 + C_1 \geq C_2 & \text{for the collection } \{\{3\},\{1,2\}\} \\ \frac{1}{2}(C_1 + C_1 + C_2) \geq C_2 & \text{for the collection } \{\{1,2\},\{1,3\},\{2,3\}\} \end{cases}$$

From this we get:

$$\begin{cases} 3C_1 & \geq C_2 \\ C_1 & \geq 0 \\ 2C_1 & \geq C_2 \end{cases}$$

So, the core is not empty if and only if:

$$0 \leq C_1 \text{ and } C_2 \leq 2C_1.$$

Notice that $C_2 < 0$ is not harming, even if it looks a little bit unrealistic. So, if we assume that C_1 and C_2 are both non-negative, the core is not empty if and only if:

$$0 \leq C_2 \leq 2C_1.$$

It has been also found that the Shapley value does not lie in the core if $C_2 > \frac{5}{3}C_1$. This is obvious if $C_2 > 2C_1$, since in such a case the core is empty. But we are left with the interesting case that for $2C_1 \geq C_2 > \frac{5}{3}C_1$ the Shapley value does not lie in the core, even if it is not empty. This is the case, for example, if $C_1 = 11$ and $C_2 = 20$.

3.2 Considerations on interpretation (b)

Following interpretation (b), we are going here and in Section 4 to perform calculations on the ex-ante game deduced by the probability distribution on the states of nature and, as consequence, on the related costs for water treatment. Of course, we must keep in mind that ex-ante costs could reasonably be very different than ex-post, since they depend on the occurrence ex-post of a particular state of nature. More precisely, under the current interpretation of the cost function in Figure 1.1, the theoretical considerations on the ex-ante cooperative game are only aimed to find a ("fair") cost sharing agreement of the cost incurred by the larger coalition ex-post.

Consider 3 players $N = \{1, 2, 3\}$ and a set of 3 states of nature $\Omega = \{\omega_1, \omega_2, \omega_3\}$.

For each state $\omega \in \Omega$ and for each player $i \in N$, values of $q(i, \omega)$ representing the amount of water to be treated are reported in the following table:

state of nature / players	ω_1	ω_2	ω_3
1	Q_1	Q_2	Q_3
2	Q_1	Q_2	Q_3
3	$2Q_1$	$2Q_2$	$2Q_3$

where $Q_1 = 2$, $Q_2 = 3$, $Q_3 = 5$ and $Q^* = 10.9$. Moreover we suppose a probability distribution on Ω such that $prob(\omega_1) = \frac{2}{5}$, $prob(\omega_2) = \frac{2}{5}$ and $prob(\omega_3) = \frac{1}{5}$. Then, by interpretation (b), the characteristic cost function for each state of nature is summarized in the following schema:

	ω_1	ω_2	ω_3	average cost
$\{1\}$	C_1	C_1	C_1	C_1
$\{2\}$	C_1	C_1	C_1	C_1
$\{3\}$	C_1	C_1	C_1	C_1
$\{1,2\}$	C_1	C_1	C_1	C_1
$\{1,3\}$	C_1	C_1	C_2	$\frac{4}{5}C_1 + \frac{1}{5}C_2 = C_1 + \frac{1}{5}(C_2 - C_1)$
$\{2,3\}$	C_1	C_1	C_2	$\frac{4}{5}C_1 + \frac{1}{5}C_2 = C_1 + \frac{1}{5}(C_2 - C_1)$
$\{1,2,3\}$	C_1	C_2	C_2	$\frac{2}{5}C_1 + \frac{3}{5}C_2 = C_1 + \frac{3}{5}(C_2 - C_1)$

For example, consider coalition $\{2, 3\}$: the cost for the treatment plant under state ω_2 is C_1 since $q(2, \omega_2) + q(3, \omega_2) = Q_2 + 2Q_2 = 9 < Q^*$, whereas under state ω_3 the cost should be C_2 since $q(2, \omega_3) + q(3, \omega_3) = Q_3 + 2Q_3 = 15 > Q^*$. Therefore, the cost for coalition $\{2, 3\}$ in the resulting (ex-ante) game should be the average cost reported in the previous table.

Now we are going to consider the resulting (ex-ante) game $\langle\{1, 2, 3\}, \bar{c}\rangle$, where $\bar{c}(S)$ is given by the average cost of $S \subseteq N$ as in the previous table. Also the average game is not concave for $C_2 > C_1$. In fact, the condition (1) is not satisfied by coalitions $S = \{1\}$ and $T = \{1, 3\}$ and $i = 2$.

The core *Core*$(\{1, 2, 3\}, \bar{c})$ of the game $\langle\{1, 2, 3\}, \bar{c}\rangle$ is given by all the vectors $(x_1, x_2, x_3) \in \mathbb{R}^3$ such that:

$$\begin{cases} x_i \leq C_1 & \forall i \in \{1,2,3\} \\ x_1 + x_2 \leq C_1 \\ x_1 + x_3 \leq C_1 + \frac{1}{5}(C_2 - C_1) \\ x_2 + x_3 \leq C_1 + \frac{1}{5}(C_2 - C_1) \\ x_1 + x_2 + x_3 = C_1 + \frac{3}{5}(C_2 - C_1) \end{cases} \quad (2)$$

Then, we can write:

$$\begin{cases} x_i \leq C_1 & \forall i \in \{1,2,3\} \\ x_1 + x_2 + x_3 - x_3 \leq C_1 \\ x_1 + x_3 + x_2 - x_2 \leq C_1 + \frac{1}{5}(C_2 - C_1) \\ x_2 + x_3 + x_1 - x_1 \leq C_1 + \frac{1}{5}(C_2 - C_1) \\ x_1 + x_2 + x_3 = C_1 + \frac{3}{5}(C_2 - C_1) \end{cases}$$

$$\begin{cases} x_i \leq C_1 & \forall i \in \{1,2,3\} \\ C_1 + \frac{3}{5}(C_2 - C_1) - x_3 \leq C_1 \\ C_1 + \frac{3}{5}(C_2 - C_1) - x_2 \leq C_1 + \frac{1}{5}(C_2 - C_1) \\ C_1 + \frac{3}{5}(C_2 - C_1) - x_1 \leq C_1 + \frac{1}{5}(C_2 - C_1) \\ x_1 + x_2 + x_3 = C_1 + \frac{3}{5}(C_2 - C_1) \end{cases}$$

$$\begin{cases} C_1 \geq x_3 \geq \frac{3}{5}(C_2 - C_1) \\ C_1 \geq x_2 \geq \frac{2}{5}(C_2 - C_1) \\ C_1 \geq x_1 \geq \frac{2}{5}(C_2 - C_1) \\ x_1 + x_2 + x_3 = C_1 + \frac{3}{5}(C_2 - C_1) \end{cases}$$

Finally
$C_1 + \frac{3}{5}(C_2 - C_1) = x_1 + x_2 + x_3 \geq \frac{7}{5}(C_2 - C_1) \Leftrightarrow C_1 \geq \frac{4}{5}(C_2 - C_1) \Leftrightarrow$ $9C_1 \geq 4C_2 \Leftrightarrow C_1 \geq \frac{4}{9}C_2$.
So if $C_2 > \frac{9}{4}C_1$ then $Core(\{1,2,3\}, \bar{c})$ is empty.

Is the Shapley value in $Core(\{1,2,3\}, \bar{c})$? Consider the following table concerning the incremental contribution of players $1, 2, 3$ to the cost of each permutation in the game $\langle\{1,2,3\}, \bar{c}\rangle$:

	1	2	3
123	C_1	0	$\frac{3}{5}(C_2 - C_1)$
132	C_1	$\frac{2}{5}(C_2 - C_1)$	$\frac{1}{5}(C_2 - C_1)$
213	0	C_1	$\frac{3}{5}(C_2 - C_1)$
231	$\frac{2}{5}(C_2 - C_1)$	C_1	$\frac{1}{5}(C_2 - C_1)$
312	$\frac{1}{5}(C_2 - C_1)$	$\frac{2}{5}(C_2 - C_1)$	C_1
321	$\frac{2}{5}(C_2 - C_1)$	$\frac{1}{5}(C_2 - C_1)$	C_1
total	$2C_1 + \frac{5}{5}(C_2 - C_1)$	$2C_1 + \frac{5}{5}(C_2 - C_1)$	$2C_1 + \frac{8}{5}(C_2 - C_1)$
$\phi_i =$	$\frac{C_1}{3} + \frac{5}{30}(C_2 - C_1)$	$\frac{C_1}{3} + \frac{5}{30}(C_2 - C_1)$	$\frac{C_1}{3} + \frac{8}{30}(C_2 - C_1)$

For individual rationality of players $1, 2, 3$:
$\phi_1 = \phi_2 = \frac{C_1}{3} + \frac{5}{30}(C_2 - C_1) \leq C_1 \Leftrightarrow \frac{5}{30}(C_2 - C_1) \leq \frac{2}{3}C_1 \Leftrightarrow 5(C_2 - C_1) \leq 20C_1$,
$\phi_3 = \frac{8}{30}(C_2 - C_1) < \frac{2}{3}C_1 \Leftrightarrow 8(C_2 - C_1) \leq 20C_1$.
For coalitional rationality for coalitions $\{1, 3\}$ and $\{2, 3\}$
$\phi_1 + \phi_3 = \phi_2 + \phi_3 = \frac{2}{3}C_1 + \frac{13}{30}(C_2 - C_1) \leq C_1 + \frac{1}{5}(C_2 - C_1) \Leftrightarrow$
$\Leftrightarrow \frac{7}{30}(C_2 - C_1) \leq \frac{1}{3}C_1 \Leftrightarrow 14(C_2 - C_1) \leq 20C_1$
and for coalition $\{1, 2\}$
$\phi_1 + \phi_2 = \frac{2}{3}C_1 + \frac{10}{30}(C_2 - C_1) \leq C_1 \Leftrightarrow \frac{10}{30}(C_2 - C_1) \leq \frac{1}{3}C_1 \Leftrightarrow$
$\Leftrightarrow 20(C_2 - C_1) \leq 20C_1 \Leftrightarrow C_2 \leq 2C_1$.

Note that if the last constraint is satisfied then also the other constraints are satisfied. Therefore, if $C_2 > 2C_1$ then the Shapley value is not in $Core(\{1, 2, 3\}, \bar{c})$.

So far our game does not take into account the attitude of the players toward risk. Will the results, and especially the core conditions change with the introduction of risk attitude of the players?

4. AN EXAMPLE OF CALCULATION OF THE CORE FOR A STOCHASTIC COOPERATIVE GAME

The model considered in this section is aimed to deal with the risk attitude of the players.

First, a preliminary introduction to stochastic cooperative games (scg) (Suijs, 1998; Suijs et al., 1999; Suijs and Borm, 1999). An scg is a tuple $\Gamma = \left(N, (A_S)_{S \subseteq N}, (X_S)_{S \subseteq N}, (\succeq_i)_{i \in N}\right)$ where

- A_S is the (finite) set of possible actions which coalition $S \subseteq N$ can take.
- A stochastic variable $X_S(a)$ corresponds to each action $a \in A_S$, that is: $X_S : A_S \to L^1(\mathbb{R})$, with $L^1(\mathbb{R})$ being the set of stochastic variables with finite expected value.
- $(\succeq_i)_{i \in N}$ is the preference profile for players in N on stochastic payoff (cost).

An allocation of $X_S(a)$ to the players in S is represented by a pair with $\sum_{i \in S} d_i \geq 0$, $\sum_{i \in S} r_i = 1$ and $r_i \geq 0$ for all players $i \in S$. Given such a pair $(d, r|a)$, agent $i \in S$ receives the stochastic cost $d_i + r_i X_S(a)$. The second part, $r_i X_S(a)$, describes the fraction of $X_S(a)$ that is allocated to player i. The first part, d_i, describes the deterministic transfer payments between the players[2]. When $d_i \leq 0$ then player i receives money while $d_i > 0$ means that this player pays money. The purpose of these transfer payments is that the players compensate each other for transfers of random costs. For example, a risk-adverse player (that "hates" uncertainty) who receives a large fraction of $X_S(a)$ can be compensated by the other players if they give him an adequate negative amount d_i. The set $Z(S)$, $S \subseteq N$, of allocations that coalition S can obtain contains all such allocations for all $X_S(a)$.

Note the relevance of indicating which action has been taken in correspondence to a given allocation. Having in mind the standard framework of a deterministic TU-game, one could object that allocations can be represented simply as repartitions of a cost (or a gain) and the actions performed do not take a relevant role in the modelling process. In the stochastic case, instead, actions are essential, as shown in the following very elementary example, where the set of possible actions represents the set of possible technologies for water treatment for players.

Example Consider a group of two players 1 and 2, that can choose between only two possible actions (i.e. technologies which can independently be used for a complete water treatment): action t_1 means "treating water by technology one" and action t_2 means "treating water by technology two". Moreover, suppose that, for budget constraints, only one technology can be implemented by the players (e.g. initial fixed costs of implementation are too high for using both of them).

Of course the cost of water treatment is conditioned, for both of the available technologies, on the state of nature which is going to occur. Figure 3 shows the cost functions related to the two technologies t_1 and t_2, when employed to treating a certain amount of water.

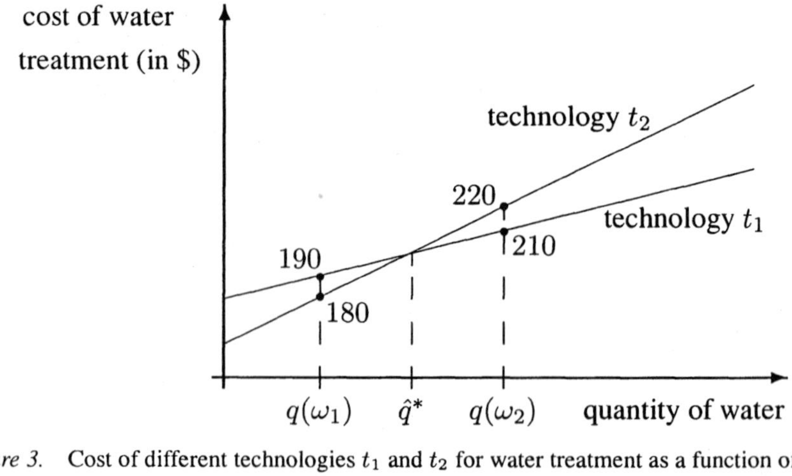

Figure 3. Cost of different technologies t_1 and t_2 for water treatment as a function of amount of water to be treated

In addition, suppose that there are only two possible states of nature: state ω_1 corresponds to a dry season and state ω_2 to a season characterized by a normal amount of rain.

Suppose that in state ω_1 (dry season) the quantity of rain water is smaller than $\hat{q}^*$, whereas in state ω_2 (normal range) the correspondent amount of water is bigger than $\hat{q}^*$. A possible situation arising from the availability of the two technologies under the two states of nature is shown in Figure 4. Points in the figure originate from possible allocations of equal (i.e. 50-50) costs between the players in each possible state of nature: realistic allocations cannot leave aside the correspondent action. In particular, an allocation allowing each player to pay 90 in state ω_1 and 105 in state ω_2 is not feasible. ■

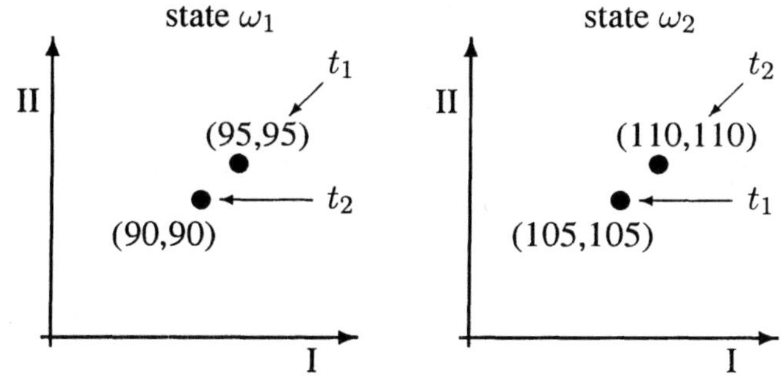

Figure 4. Consequences (in terms of cost allocations between player 1 and 2) of actions t_1 and t_2 in each possible state

Of course, it is also possible to define counterparts of TU-game solution concepts in the framework of stochastic cooperative games. For example the definition of core allocation is the following (Suijs et al., 1999):

Definition 1*[Core Allocation] An allocation* $(d, r|a) \in Z(N)$ *is a core allocation if* $\nexists\, S \subseteq N$ *and* $(\hat{d}, \hat{r}|\hat{a}) \in Z(N)$ *s.t.* $(\hat{d}, \hat{r}|\hat{a})_i \succ_i (d, r|a)_i$ *for each* $i \in S$.

Therefore the core is the set of all the core allocations, that is:
$Core(\Gamma) = \{(d, r|a) \in Z(N) : \nexists(\tilde{d}, \tilde{r}|\tilde{a}) \in Z(S) \text{ s.t. } (\tilde{d}, \tilde{r}|\tilde{a})_i \succ_i (d, r|a)_i \; \forall\, i \in S\}$.

Since we do not see which different actions players could take in the interpretation (b) of the cost function, then we assume that each coalition actually has the possibility to choose a unique action (in other words we assume that to each coalition corresponds a set of precisely one action).

Moreover, at the present we assume that there are no possibilities of ex-ante deterministic transfer payments. So we suppose $d_i = 0$ for each $i \in N$. On the other hand, no particular computational difficulties should arise if we would be willing to take into account both these deterministic transfers and different actions.

Even in the very simple setting that we have considered (that is: a *unique* action available to all the coalitions and no "ex-ante" payments allowed), it is possible to see that *under risk aversion* there is room for situations, which show new phenomena with respect to the "expected cost" TU-game (i.e., for risk neutral players).

Consider the following situation:

$$N = \{1, 2, 3\}, \quad \Omega = \{\omega_1, \omega_2\}, \quad p(\omega_1) = 1/3, \quad p(\omega_2) = 2/3.$$

The costs for the various coalitions are the following:

$$\begin{array}{lcl} c(\{i\}) & = & 0 \text{ in any state} \\ c(\{N\}) & = & 2 + \delta \text{ in any state} \\ c(\{1,2\}) & = & \left\{ \begin{array}{ll} 1 & \text{in state } \omega_1 \\ 3/2 & \text{in state } \omega_2 \end{array} \right. \\ c(\{1,3\}) & = & 4/3 \\ c(\{2,3\}) & = & 4/3 \end{array}$$

From these data we can get the "expected cost" game, where the values for the coalitions are given by the expected costs. The expected value of $c(\{1, 2\})$ is $1 \cdot 1/3 + 3/2 \cdot 2/3 = 4/3$. For other coalitions, the costs are sure.

Necessary and Sufficient condition for nonemptyness of a core of this "expected cost" game is (Bondareva-Shapley theorem):

$$1/2 \cdot 4/3 + 1/2 \cdot 4/3 + 1/2 \cdot 4/3 \geq 2 + \delta$$

That is: $\delta \leq 0$.

In particular, if $\delta = 0$, the core is not empty (and $(2/3, 2/3, 2/3)$ is easily checked to be a unique element of the core).

In general, if $\delta > 0$, the core for the "expected cost game" is empty. But, if players 1 and 2 are enough risk adverse (in terms of the Arrow – Pratt coefficients; cf. Arrow, 1995; Pratt, 1964), the core of the stochastic game can be *non empty*.

Let us consider the allocation

$$\left(\frac{2}{3} + \delta, \frac{2}{3} + \delta, \frac{2}{3} - \delta\right).$$

This allocation is feasible:

$$\frac{2}{3} + \delta + \frac{2}{3} + \delta + \frac{2}{3} - \delta = 2 + \delta = c(N).$$

To see that (under suitable conditions) this allocation is in the core, we consider the coalition $\{1, 2\}$ (later we shall consider also the other coalitions).

The coalition can "object" to the proposed allocation in case there is a Pareto efficient division of the stochastic cost for $\{1, 2\}$, such that both prefer it to what they should get from the proposed allocation.

We assume that *players* 1 *and* 2 *have identical vN-M preferences*, in particular the same risk attitude, that we shall assume to be constant (as measured by the Arrow-Pratt coefficient). In such a case, taking into account Suijs (1998), in particular Proposition 3.1, it is enough to refer to an even (in equal parts) division of the costs for $\{1, 2\}$.

To realize this, remember that we must find the Pareto allocations for $\{1, 2\}$. But these clearly correspond to a division of the (stochastic) costs of the coalition among the two players by means of r_1, r_2, with $r_i \geq 0$ and $r_1 + r_2 = 1$. Due to the fact that the two players are identical, the *optimal division* to support an objection against the proposed allocation of $c(N)$ is the equal division.

So, we compare (we do the calculations for player 1, but player 2 is identical to player 1 so the same calculations apply) the *certain* cost $\frac{2}{3} + \delta$ with the stochastic cost: $\frac{1}{2} \cdot 1$ in state ω_1 and $\frac{1}{2} \cdot \frac{3}{2}$ in state ω_2.

Due to the concavity of the utility function for 1, it is possible that:

$$u_1\left(-\frac{2}{3} - \delta\right) \geq \frac{1}{3}u_1\left(-\frac{1}{2}\right) + \frac{2}{3}u_1\left(-\frac{3}{4}\right)$$

(which is equivalent to $\kappa \leq -\frac{2}{3} - \delta$, where κ is the certainty equivalent for player 1 of the lottery L which gives $-1/2$ with probability $1/3$ and $-3/4$ with probability $2/3$. Notice that, due to risk aversion, we have that $\kappa < -2/3$).

So, if the above condition is satisfied, the proposed allocation

$$\left(\frac{2}{3} + \delta, \frac{2}{3} + \delta, \frac{2}{3} - \delta\right)$$

cannot be "improved upon" by $\{1, 2\}$.

Let's check for the coalition $\{2, 3\}$. Since costs are deterministic, the condition for "not being improved upon" is:

$$c(\{2, 3\}) = \frac{4}{3} \geq \frac{2}{3} + \delta + \frac{2}{3} - \delta,$$

which is obviously satisfied.

And the same calculations apply for the coalition $\{1, 3\}$.

5. CONCLUSION

Application of cooperative game theory to water resource problems in the literature was done mainly under deterministic supply pattern. Not only this assumption is not realistic, but it also leads to solutions of such problems that are not in the core. Furthermore, with increased likelihood of climate change, the extreme events in the water supply chain will become more and more frequent and prominent, and thus more relevant. We attempted to apply a stochastic game theory framework to demonstrate how stochastic considerations may affect the nature of the solution, depending on the attitude of the players toward risk and the nature of the (possibly non concave) cost function of the water project.

We show that in the case of a deterministic game, for both types of the cost function, the Shapley Value is not in the core. Taking into account stochastic costs, in Section 4 we have shown that it can happen that risk averse players end up in a solution that lays in the core, while risk indifferent players end up in a solution that lays outside of the core.

Our major conclusion from the analysis in the chapter is that stochastic games can provide a solid basis for the analysis of water games in various sectors needing joint facilities, such as water treatment, water storage, and irrigation. Since the problem of ex-ante planning of both the investment and the operation of the joint facility is so crucial for the future sustainability of the facility and externalities derived from its operation (such as the case of the Cahora Bassa), we would suggest to water project planners to include stochastic considerations at the initial planning of each project.

Notes

1. For a general introduction to cooperative Game Theory see Owen (1995).
2. Notice that Suijs et al. (1999), proposed also a different condition on the deterministic part, that is: $\sum_{i \in S} d_i = \mathbb{E}(X_S(a))$.

References

Arrow, K.J., 1965. 'The Theory of Risk Aversion', *Lecture 2 in Aspects of the Theory of Risk-Bearing (Yrjo Jahnsson Lectures)*, Yrjo Jahnssonin Saatio, Helsinki.

Bondareva, 0., 1963. 'Certain applications of the methods of linear programming to the theory of cooperative games' (In Russian), *Problemy Kibernetiki*, 10, 119-139.

Dinar, A., M. B. Campbell, and D. Zilberman, 1992. 'Adoption of improved irrigation and drainage reduction technologies under limiting environmental conditions', *Environmental & Resource Economics*, 2, 373-398.

Dinar, A. and R.E. Howitt, 1997. 'Mechanisms for allocation of environmental control cost: empirical tests of acceptability and stability', *Journal of Environmental Management*, 49, 183-203.

Dinar, A. and D. Zilberman, 1991. 'The economics of resource-conservation, pollution-reduction technology selection: The case of irrigation water', *Resources and Energy*, 13, 323-348.

Dinar, A., D. Yaron and Y. Kannai, 1986. 'Sharing regional cooperative gains from reusing effluent for irrigation', *Water Resources Research*, 22(2), 339-344.

Gately, D., 1974. 'Sharing the gains from regional cooperation: A game theoretic application to planning investment in electric power', *International Economic Review*, 15(1), 195-208.

Loehman, E., J. Orlando, J. Tschirhart and A. Whinston, 1979. 'Cost allocation for a regional wastewater treatment system', *Water Resources Research*, 15(2), 193-202.

Moretti, S., H. Norde, K.H. Pham Do and S. Tijs, 2002. 'Connection problems in mountains and monotonic allocation schemes', *Top*, 10, 83-99.

Owen, G., 1995. 'Game Theory', *Academic Press*, Third edition.

Pratt, J.W., 1964. 'Risk aversion in the small and in the large', *Econometrica*, 32, 122-136.

Rogers, P., 1993. 'The value of cooperation in resolving international river basin disputes', *Natural Resources Forum*, 17, 117-131.

Shapley, L.S., 1967. 'On balanced sets and cores', *Naval research logistic quarterly*, 14, 453-460.

Straffin, P. D. and J. P. Heaney, 1981. 'Game theory and the Tennessee Valley authority', *International Journal of Game Theory*, 10(1), 35-43.

Suijs, J., 1998. 'Cooperative Decision Making in a Stochastic Environment', PhD Thesis, University of Tilburg.

Suijs, J., 2003. 'Cost allocation in spanning network enterprises with stochastic connection costs', *Games and Economic Behavior*, 42, 156-171.

Suijs, J. and P. Borm, 1999. 'Cooperative games: Superadditivity, convexity, and certainty equivalents', *Games and Economic Behavior*, 27, 331-345.

Suijs, J., P. Borm and A. De Waegenaere, 1998. 'Stochastic cooperative games in insurance', *Insurance: Mathematics & Economics*, 22, 209-228.

Suijs, J., P. Borm, A. De Waegenaere and S. Tijs, 1999. 'Cooperative games with stochastic payoffs', *European Journal of Operational Research*, 113, 193-205.

Suzuki, M. and M. Nakayama, 1976. 'The cost assignment of cooperative water resource development -A game theoretic approach', *Management Science*, 22, 1081-1086.

Tanimoto, K., N. Okada and H. Tatano, 2000. 'Mutual Risk Trade through Stochastic Cost Allocation in the Recovery Project from Disaster', *Proc. of Second Euro Conference on Global Change and Catastrophe Risk Management*, http://www.iiasa.ac.at/Research/ RMP/july2000/.

Chapter 2

INCENTIVES AND THE SEARCH FOR UNKNOWN RESOURCES SUCH AS WATER*

Jean-Jacques Laffont
Université Toulouse and CNRS, France, University of Southern California, U.S.A.

François Salanié
Université Toulouse (INRA & LERNA), France

1. INTRODUCTION

The delegation of discovery tasks is quite common. Researchers are funded by public and private organizations for discovering new theorems, new computer algorithms, new engineering processes. Multinational companies are often delegated the search and exploitation of oil fields or other natural resources such as water. Communities often own in common resources of unknown magnitudes like water, wood, plants, fruits, game in forests or fish in rivers and oceans. In general, communities specialize some individuals (hunters, fishermen) to look for these resources.

Hence most of the R&D efforts are delegated by principals to agents through labor contracts. The R&D literature has well taken into account the randomness of discoveries and the need to structure contracts for giving proper incentives to the agents in charge of R&D tasks. One essential feature of the discovery process which has not been taken into account is that, almost by definition, the nature or size of the discovery is private information of the agent who makes the

*This paper was prepared for the talks given at the University of California at Berkeley and Davis and sponsored by the Giannini Foundation and for the book Frontiers in Water Resource Economics. A first draft of this paper was prepared by Jean-Jacques Laffont before his death on 1 May 2004. The paper was substantially revised by François Salanié, with the help of the comments of the referee, in the Spring of 2005.

discovery. This may lead to opportunistic behavior of agents. Taking advantage of the "no slavery" conditions of labor contracts, researchers may hide their true discovery, renege on their labor contract and exploit their discovery outside the principal-agent relationship. Companies which are delegated by countries the search for natural resources may take advantage of the lack of technical expertise of these countries to hide some of their discoveries or even flee the country with their discovery. Similarly, hunters or fishermen may hide some of their findings and consume them secretly.

The common structure of these examples is a principal-agent problem in which a principal delegates to an agent the search for a resource of unknown size, and the outcome of the search is partially nonverifiable, and contracts are imperfect. In this paper we study this contractual problem by modelling the imperfection of contracts as an imperfection in the enforcement of contracts combined with limited liability constraints, noticeable characteristics of Less Developed Countries.

Let us illustrate our main findings within the following example. A company is in charge of finding water in the mountains. Part of the discovered water must be piped to the downstream town, the remaining being consumed locally. This sharing problem is made difficult because only the company knows the size of the discovery. Moreover, when the discovery is big enough the company may want to quit the principal-agent relationship to exploit the discovery alone, for example through selling water on a black market. To avoid this outcome, the principal (here the town) has to reward a company announcing a big discovery, thus creating an incentive to lie for companies whose discovery is of a small size. The optimal contract is shown to include unusual distortions: companies with big discoveries should optimally be asked to provide more water to the town than is socially optimal, so as to deter companies with small discoveries to mimic them.

Moreover, in such a regime the principal turns out to be better off when the size of the discovery is smaller. This creates perverse incentives, since the principal may want to discourage the search effort exerted by the agent. In an extension of the model including an endogenous search effort, we show that this is indeed the case. The optimal contract discourages the search effort by rewarding companies announcing small discoveries.

These findings rely on the imperfection of contracts, that is, the town cannot deter a company with a big discovery to turn to the black market if it finds it profitable. The key problem is thus a question of enforcement. We show that with better enforcement more natural incentives are restored. We also discuss by means of examples how the perverse incentives may appear or not, depending on precise properties of the agent's payoffs.

This paper is related to several strands of the economic literature. The first one deals with principal-agent models with interim status-quo payoffs which are state dependent (Lewis and Sappington, 1989; Laffont and Tirole, 1990; Maggi and Rodriguez-Clare, 1995; Jullien, 2000). In our analysis the imperfection of enforcement will yield such interim constraints despite the fact that contracts are signed ex ante. The second one is the law and economics literature initiated by Becker (1968), Stigler (1970), Becker and Stigler (1974) about the imperfect enforcement of laws, rules and contracts. Even though it would be desirable to derive this imperfection from explicit transaction costs of the enforcement mechanism we will adopt a rather ad hoc formulation as in this literature. The third strand of the related literature concerns the manipulation of endowments in mechanism design (Postlewaite, 1979; Hurwicz et al., 1982; Green and Laffont, 1986). However, for reasons to be given below we will not use the major insight of these papers which is to argue that one can only lie downward about endowments.

Section 2 sets up the model and derives the optimal contract under complete information. Section 3 derives the precise structure of the optimal contract under asymmetric information when enforcement is imperfect. This structure implies in Section 4 that the principal has the perverse incentive to discourage the agent's effort for high quality discovery. Section 5 shows how an improvement of enforcement may restore the optimality of contract and reverse these perverse incentives. Various extensions are discussed in Section 6. It is shown in particular that the details of enforcement imperfections matter a lot for the qualitative features of the optimal contract. Concluding comments are gathered in Section 7.

2. THE MODEL

We consider a principal-agent relationship in which a principal delegates to an agent the search for a resource of unknown magnitude. For reference we consider in this section the case where the amount θ of resource is known. This amount must be shared between the principal and the agent.

The principal's utility function is $u(q) - t$, where u is concave, q is the quantity of the resource obtained by the principal, and t is the monetary payment made by the principal to the agent. By symmetry the agent's utility function is $u(\theta - q) + t$. These surpluses may be interpreted as the revenues from selling the resource to final consumers. We normalize these revenues so that $u(0) = 0$.

Under complete information the principal maximizes his utility under the participation constraint of the agent

$$u(\theta - q) + t \geq 0.$$

For each θ the optimal solution $(q^*(\theta), t^*(\theta))$ is characterized by equal sharing of the good $q^*(\theta) = \theta/2$, with $t^*(\theta) = -u\left(\frac{\theta}{2}\right)$.

3. OPTIMAL CONTRACT UNDER ASYMMETRIC INFORMATION

The amount discovered θ is now private information of the agent. θ can take two values $\underline{\theta}$ and $\bar{\theta}$ with respective probabilities $1 - \nu$ and ν. Let $\Delta\theta$ denote the difference $\bar{\theta} - \underline{\theta}$. From the revelation principle we can focus on incentive compatible menus of contracts $\{\underline{t}, \underline{q}; \bar{t}, \bar{q}\}$, verifying the following two incentive compatibility constraints

$$u(\bar{\theta} - \bar{q}) + \bar{t} \geq u(\bar{\theta} - \underline{q}) + \underline{t} \quad (1)$$

$$u(\underline{\theta} - \underline{q}) + \underline{t} \geq u(\underline{\theta} - \bar{q}) + \bar{t}. \quad (2)$$

The contract between the principal and the agent is signed at the ex ante stage and subject to the agent's ex ante participation constraint:

$$\nu(u(\bar{\theta} - \bar{q}) + \bar{t}) + (1 - \nu)(u(\underline{\theta} - \underline{q}) + \underline{t}) \geq 0. \quad (3)$$

Furthermore, we assume that they are some enforcement difficulties originating in the legal environment. Firstly, the agent is protected by limited liability, and cannot end up with a negative payoff. Secondly, we suppose that at any time the agent can quit the principal-agent relationship[1] and exploit the discovery by himself. In such a case courts can only impose that the amount $\underline{\theta}$ is left to the Principal. Hence, if the agent discovers a high amount $\bar{\theta}$, he can disappear with the amount $\Delta\theta$ of the resource, thus getting the payoff $w \equiv u(\Delta\theta)$. This leads to the additional enforcement constraints:

$$\underline{U} = u(\underline{\theta} - \underline{q}) + \underline{t} \geq 0 \quad (4)$$

$$\bar{U} = u(\bar{\theta} - \bar{q}) + \bar{t} \geq w. \quad (5)$$

The principal's best menu of contracts maximizes, under (1) to (5) the following objective function:[2]

$$\nu(u(\bar{q}) - \bar{t}) + (1 - \nu)(u(\underline{q}) - \underline{t}). \quad (6)$$

In the absence of enforcement constraints, the principal facing a risk neutral agent with an ex ante participation constraint would achieve his first-best.[3] Note also that (3) is implied by (4) and (5), so that the only constraints to consider are (1) (2) (4) (5). The problem then becomes similar to a principal agent problem with interim participation constraints and type-dependent status-quo utility levels for the agent.[4]

Appendix A derives the shape of the optimal contract for arbitrary values for w. It is shown that $\underline{q}$ must be *distorted downward* (below the first-best level $q^*(\underline{\theta}) = \underline{\theta}/2$), while $\bar{q}$ must be *distorted upward* (above $q^*(\bar{\theta}) = \bar{\theta}/2$). The point we would like to underline is the following. For $w = u(\Delta\theta)$, the usual result that the incentive constraint (1) and the enforcement constraint (4) are binding does not hold anymore. Indeed the rent of the "good" type $\bar{\theta}$ would be equal to

$$u(\bar{\theta}-\bar{q})+\bar{t}=u(\bar{\theta}-\underline{q})+\underline{t}=u(\bar{\theta}-\underline{q})-u(\underline{\theta}-\underline{q})$$

$$=u(\Delta\theta+\underline{\theta}-\underline{q})-u(\underline{\theta}-\underline{q})<u(\Delta\theta)=w$$

where the last inequality comes from the concavity of u and the fact that $\underline{q}<\underline{\theta}$. Thus (5) would not hold. We are then led to distinguish between three regimes.

Regime 1: this regime obtains when only the enforcement constraints are binding. Substituting these constraints in the principal's objective function and maximizing with respect to $\underline{q}, \bar{q}$, we obtain efficient sharing. In this case, the principal's expected welfare is

$$W_1=2\nu u\left(\frac{\bar{\theta}}{2}\right)+2(1-\nu)u\left(\frac{\underline{\theta}}{2}\right)-\nu u(\Delta\theta).$$

With respect to the first best, the principal loses only $\nu u(\Delta\theta)$. Such a contract is optimal as soon as it verifies the constraint (2), or equivalently if $\Delta\theta\geq\underline{\theta}$.

In the other two regimes we have countervailing incentives, i.e., it is the incentive constraint (2) of the "bad" type $\underline{\theta}$ which is binding.

Regime 3: the enforcement constraint of the good type $\bar{\theta}$ and the incentive constraint of the bad type $\underline{\theta}$ are binding. Optimizing quantities we get:

$$\underline{q}_3 = q^*(\underline{\theta})=\frac{\underline{\theta}}{2} \tag{7}$$

$$u'(\bar{q}_3) = u'(\bar{\theta}-\bar{q}_3)-\frac{(1-\nu)}{\nu}\left(u'(\underline{\theta}-\bar{q}_3)-u'(\bar{\theta}-\bar{q}_3)\right). \tag{8}$$

The principal's expected welfare is then

$$\begin{aligned} W_3 &= \nu(u(\bar{q}_3)+u(\bar{\theta}-\bar{q}_3))-\nu u(\Delta\theta) \\ &+2(1-\nu)u\left(\frac{\underline{\theta}}{2}\right)-(1-\nu)\left(u(\Delta\theta)-(u(\bar{\theta}-\bar{q})-u(\underline{\theta}-\underline{q}))\right). \end{aligned}$$

However, this case is valid only if the $\underline{\theta}$ agent's participation constraint is satisfied :

$$u(\Delta\theta)-(u(\bar{\theta}-\bar{q}_3)-u(\underline{\theta}-\bar{q}_3))\geq 0$$

i.e., if $\bar{q}_3\leq\underline{\theta}$.

If $\bar{q}_3>\underline{\theta}$, we have **Regime 2** which connects Regimes 1 and 3. In Regime 2, $\bar{q}_2=\underline{\theta}$ and $\underline{q}_2$ still equates $\frac{\underline{\theta}}{2}$ with both enforcement constraints binding and defining the transfers. Then, the principal's expected welfare is

$$W_2 = \nu u(\underline{\theta}) + 2(1-\nu)u\left(\frac{\underline{\theta}}{2}\right).$$

Summarizing we have:

PROPOSITION 1: *Suppose that* $w = u(\Delta\theta)$. *The optimal menu of contracts entails:*

i) Efficient sharing if the asymmetry of information is large enough ($\Delta\theta > \underline{\theta}$).

ii) Countervailing incentives and ***upward*** *distortions of the quantity allocated to the* $\bar{\theta}$*-agent, otherwise.*

For the example $u(x) = x - \frac{x^2}{2}$ we give in Figure 1 the typical profile of quantity allocated to the $\bar{\theta}$-agent.

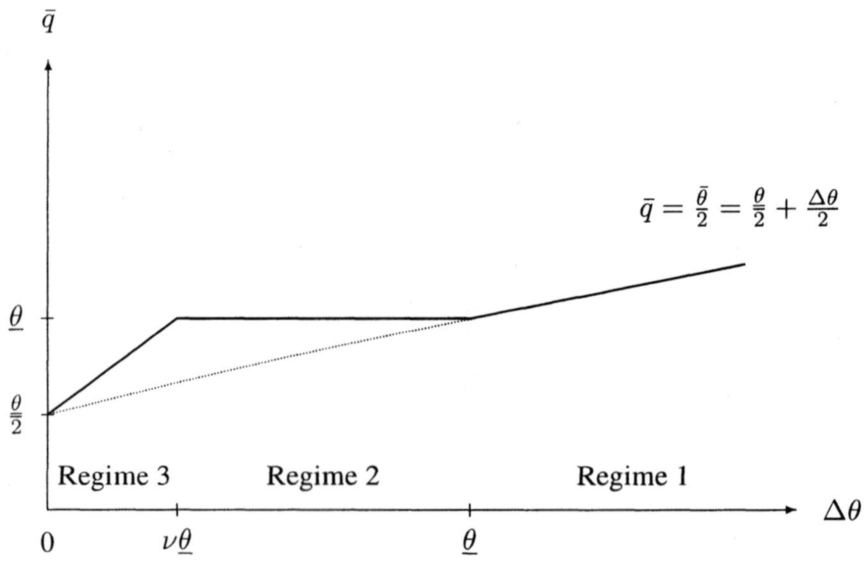

Figure 1. Typical profile of quantity allocated to the $\bar{\theta}$-agent

So far, the main implication of enforcement constraints is that they may yield unexpected distortions of the optimal contract. For w high enough (and in particular for $w = u(\Delta\theta)$), the enforcement constraint (5) requires that the principal offers type $\underline{\theta}$ a contract so favorable that it becomes attractive to the $\underline{\theta}$-type. This calls for a distortion of the $\bar{\theta}$-contract aimed at avoiding this mimicking behavior, namely an *upward* distortion (with respect to the first-best) in the quantity that $\bar{\theta}$ must provide. In the next section, we shall see that these characteristics of the optimal contract have surprising effects on the principal's incentives to favor large discoveries.

4. INCENTIVES FOR DISCOVERY EFFORT

One consequence of the derivations above is that the Principal may be better off when the discovery is small. Indeed define the Principal's payoffs as

$$\underline{V} = u(\underline{q}) - \underline{t} \qquad \bar{V} = u(\bar{q}) - \bar{t}.$$

In regimes 2 and 3, the incentive constraint (2) is binding. Replacing we get

$$\bar{V} - \underline{V} = u(\bar{q}) + u(\underline{\theta} - \bar{q}) - u(\underline{q}) - u(\underline{\theta} - \underline{q})$$

which is easily shown to be negative.[5] This shows that the Principal gets a higher payoff when the discovery is small.

The same result holds also in Regime 1, where the welfare obtained by the principal is

$$W_1(\nu) \equiv 2\left(\nu u\left(\frac{\bar{\theta}}{2}\right) + (1-\nu)u\left(\frac{\underline{\theta}}{2}\right)\right) - \nu u(\Delta\theta)$$

which is decreasing in ν from the concavity of u. Once more, the Principal gets a higher payoff when the discovery is small. This surprising result calls for a precise study of the agent's search effort, and of the incentives for effort given by the contract.

So far, the agent was unable to affect by his own behavior the probability distribution of the discovery size. We assume now that by exerting an effort which costs him ψ the agent increases the probability of a $\bar{\theta}$-discovery from ν_0 to $\nu_1 > \nu_0$ and let $\Delta\nu = \nu_1 - \nu_0$.

Because intuitively the principal wishes to discourage the agent from making an effort, let us first solve our program under the constraint that the agent exerts no effort:

$$\max_{(\bar{q},\bar{t};\underline{q},\underline{t})} \nu_0(u(\bar{q}) - \bar{t}) + (1-\nu_0)(u(\underline{q}) - \underline{t})$$

under (1)(2)(4)(5) and

$$\begin{aligned}\nu_0(u(\bar{\theta} - \bar{q}) + \bar{t}) + (1-\nu_0)(u(\underline{\theta} - \underline{q}) + \underline{t}) \\ \geq \nu_1(u(\bar{\theta} - \bar{q}) + \bar{t}) + (1-\nu_1)(u(\underline{\theta} - \underline{q}) + \underline{t}) - \psi. \qquad (9)\end{aligned}$$

This last constraint can be rewritten as

$$\bar{U} - \underline{U} \leq \frac{\psi}{\Delta\nu}.$$

In Regimes 1, 2 and 3, the left-hand-side was anyway less than $w = u(\Delta\theta)$. We therefore obtain that if $u(\Delta\theta) < \frac{\psi}{\Delta\nu}$, the principal does not need to distort the solution found in the preceding section: the rent differential is low compared to the cost of effort. Then, the principal obtains the expected welfare $W_1(\nu_0)$.

In Appendix B, we show that when $\Delta\theta$ increases, the principal increases the rent $\underline{U}$ of the $\underline{\theta}$-type to still discourage effort. For $\Delta\theta$ large, the principal gives up discouraging the agent from exerting effort and obtains the expected welfare $W_1(\nu_1)$.

We summarize these results by the following proposition. Figure 2 illustrates the impact of $\Delta\theta$ on the Principal's payoff.

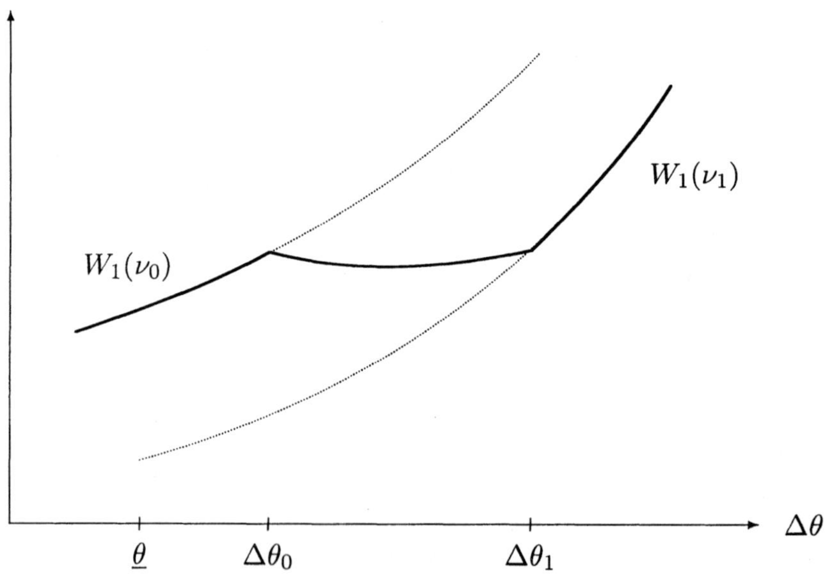

Figure 2. Impact of $\Delta\theta$ on the Principal's payoff

PROPOSITION 2: *Suppose that $w = u(\Delta\theta)$. The principal will structure incentives in order to discourage effort. This is done without cost for small $\Delta\theta$. For large $\Delta\theta$ he gives up discouraging effort. For intermediary values of $\Delta\theta$ discouraging effort is costly and requires rewarding more the $\underline{\theta}$-agent.*

The striking consequence of the enforcement constraint is that the principal has the incentive to discourage effort. If possible he will even make difficult for the agent to increase his probability of a high discovery, by denying for example proper equipment. He will also structure incentive payments so that effort is discouraged, unless it is too costly. Not only is weak enforcement calling for distortions in general and is costly to the principal. It can set up very wrong incentives for progress in society.

The intuition for this result is simply that the principal benefits more from a $\underline{\theta}$-agent than from a $\bar{\theta}$-agent because of the cost of the enforcement constraint, as was shown at the beginning of this Section. In the next section we show how an improvement of enforcement may take the relationship out of the vicious circle emphasized above.

5. ENDOGENOUS ENFORCEMENT

Suppose now that with an ex ante expense of $c(p)$ the principal can improve enforcement. More specifically, with probability p the agent is set at the zero utility level[6] and with probability $1-p$ he escapes. Let us assume that $c(\cdot)$ is convex with the Inada conditions $c(0) = 0$, $c'(0) = 0$, $\lim_{p\to 1} c'(p) = \infty$.

The ex post enforcement constraint (5) becomes

$$\bar{U} = u(\bar{\theta} - \bar{q}) + \bar{t} \geq (1-p)u(\Delta\theta). \tag{10}$$

The principal's problem can be rewritten

$$\max_{\{\bar{q},\bar{t};\underline{q},\underline{t};p\}} \nu(u(\bar{q}) - \bar{t}) + (1-\nu)(u(\underline{q}) - \underline{t}) - c(p)$$

s.t. (1) (2) (4) (10).

Thanks to enforcement expenditures it is now possible to have Regime 0, for which the $\bar{\theta}$-type incentive constraint and the $\underline{\theta}$-participation constraint are binding. Actually, it is never worth spending enforcement resources to make constraint (10) strict. So, in Regime 0, constraints (1) (3) (10) are binding. Substituting the transfers from (1) and (3) into the principal's objective function leads to:

$$\begin{aligned} \max_{\{\bar{q},\underline{q},p\}} \quad & \nu\left(u(\bar{q}) + u(\bar{\theta} - \bar{q}) - (u(\bar{\theta} - \underline{q}) - (\underline{\theta} - \underline{q}))\right) \\ & +(1-\nu)(u(\underline{q}) + u(\underline{\theta} - \underline{q})) - c(p) \end{aligned}$$

s.t.

$$u(\bar{\theta} - \underline{q}) - u(\underline{\theta} - \underline{q}) - (1-p)u(\Delta\theta) = 0.$$

The optimal solution is:

$$\bar{q}_0 = \frac{\bar{\theta}}{2} \tag{11}$$

$$u'(\underline{q}_0) = u'(\underline{\theta} - \underline{q}_0) - \frac{\left(\nu - \frac{c'(p_0)}{u(\Delta\theta)}\right)}{1-\nu}\left(u'(\bar{\theta} - \underline{q}_0) - u'(\underline{\theta} - \underline{q}_0)\right) \tag{12}$$

$$u(\bar{\theta} - \underline{q}_0) - u(\underline{\theta} - \underline{q}_0) = (1-p_0)u(\Delta\theta). \tag{13}$$

This solution holds as long as the $\underline{\theta}$-agent's incentive constraint is not binding (i.e., for $\bar{\theta} > 2\underline{\theta}$) and as long as $c'(p_0) < \nu u(\Delta\theta)$. We observe that the enforcement cost leads to a smaller distortion in $\underline{q}_0$. Indeed, when $\underline{q}_0$ is decreased to decrease the $\bar{\theta}$-agent's information rent one must take into account the addition enforcement cost due to the higher p_0 needed to maintain (13).

When $\Delta\theta$ decreases and p_0 reaches p_1 defined by $c'(p_0) = \nu u(\Delta\theta)$, the $\bar{\theta}$-agent's incentive constraint becomes slack and we obtain Regime 1 with efficient sharing $p = p_1$ and transfers $\underline{t} = -u\left(\frac{\underline{\theta}}{2}\right)$, $\bar{t} = -u\left(\frac{\bar{\theta}}{2}\right) + (1 - p_1)u(\Delta\theta)$. When $\bar{\theta} < 2\underline{\theta}$, we have also two regimes. In Regime 3, only the participation constraint of type $\bar{\theta}$ and the incentive constraint of type $\underline{\theta}$ are binding. We obtain immediately

$$\underline{q}_3 = \frac{\underline{\theta}}{2} \tag{14}$$

$$u'(\bar{q}_3) = u'(\bar{\theta} - \bar{q}_3) - \frac{(1-\nu)}{\nu}\left(u'(\underline{\theta} - \bar{q}_3) - u'(\bar{\theta} - \bar{q}_3)\right) \tag{15}$$

$$c'(p_3) = u(\Delta\theta). \tag{16}$$

For $\Delta\theta$ higher p is adjusted so that

$$(1-p)u(\Delta\theta) = u(\bar{\theta} - \bar{q}) - u(\underline{\theta} - \underline{q})$$

and we obtain Regime 2 characterized by

$$\underline{q}_2 = \frac{\underline{\theta}}{2} \tag{17}$$

$$u'(\bar{q}_2) = u'(\bar{\theta} - \bar{q}_2) + \frac{\left(1 - \nu - \frac{c'(p_2) - u(\Delta\theta)}{u(\Delta\theta)}\right)}{\nu} \cdot \left(u'(\bar{\theta} - \bar{q}_2) - u'(\underline{\theta} - \bar{q}_2)\right) \tag{18}$$

$$(1-p_2)u(\Delta\theta) = u(\bar{\theta} - \bar{q}_2) - u(\underline{\theta} - \underline{q}_2). \tag{19}$$

Putting together these results we describe in Figure 3 the profile of optimal enforcement levels. For the quadratic example we obtain in Figure 4 the quantity profiles.

When better enforcement produces Regime 0, the principal's welfare becomes increasing in ν and the principal has now incentives to induce effort. For a given level of ψ they may not want to pay the price for it, but as $\Delta\theta$ increases they will indeed structure the incentives to induce effort.

Summarizing we have:

PROPOSITION 3: *When the cost of better enforcement is low enough the principal becomes interested in increasing ν for $\Delta\theta$ large enough. He will then structure incentives to induce effort.*

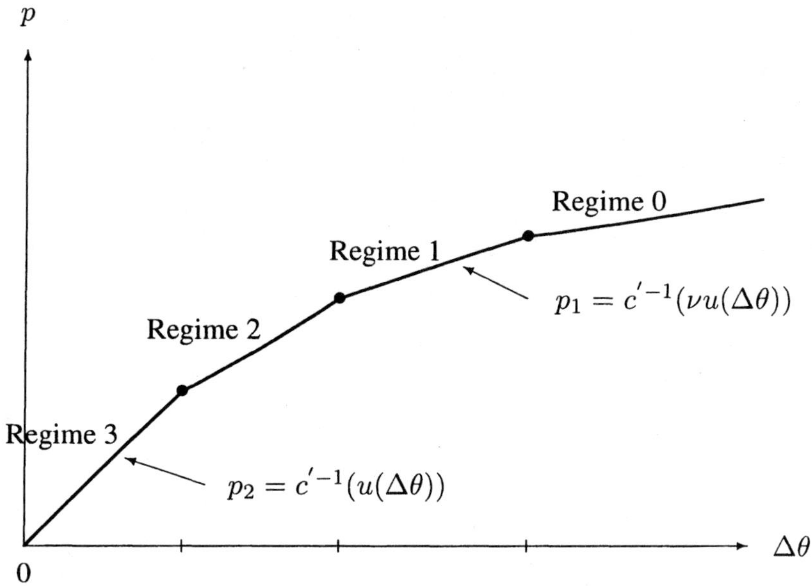

Figure 3. Profile of optimal enforcement levels

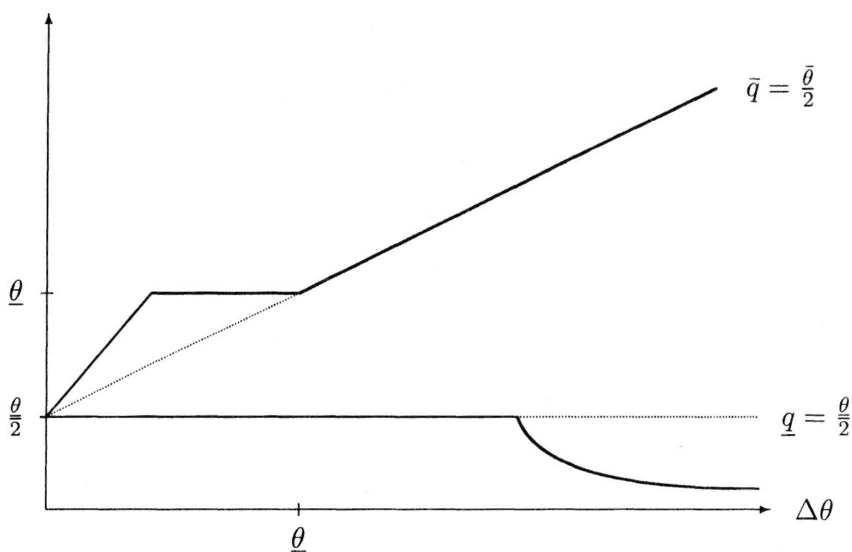

Figure 4. Quantity profiles for the quadratic example

Note that better enforcement and lower powered incentives for production are complement instruments. Indeed, better enforcement creates higher incentives for discovery effort. This in turn creates a higher probability of having a $\bar{\theta}$-agent, therefore a higher desire to limit his information rent through a lower powered incentives for production.

6. EXTENSIONS

One could argue that the agent can hide some of his discovery, but cannot claim he has discovered more than he has done in reality, because he must be able to show it. In other words, he can lie downward but not upward.[7]

In this case, there is no incentive constraint for the $\underline{\theta}$-type, and the only binding constraints are the participation constraints with efficient sharing of the discovered resource.

However, there are cases where such a strategy of asking the agent to exhibit his discovery is not possible. If it is a discovery like a computer program, just showing it provides the resource to the principal. If it is a farmer upstream on a river the water he let flow downstream cannot be recaptured. Also the agent may have hidden resources which enable him to mimic the behavior of the $\bar{\theta}$-type. The model we have studied is designed to fit those situations. The next paragraphs offer some extensions.

6.1 Modelling Imperfect Enforcement

It turns out that the details of imperfect enforcement play a crucial role in determining the structure of the optimal contract and its impact on the incentives for discovery effort.

Let us denote more generally $w(\theta)$ the outside opportunity of type θ when he reneges on the contract. If we still normalize $w(\underline{\theta}) = 0$ we have the "normal" case of Regime 0 if

$$u(\bar{\theta} - \underline{q}^{SB}) - u(\underline{\theta} - \underline{q}^{SB}) > w(\bar{\theta})$$

where $\underline{q}^{SB}$ is determined by

$$u'(\underline{q}^{SB}) = u'(\underline{\theta} - \underline{q}^{SB}) - \frac{\nu}{1-\nu}\left(u'(\bar{\theta} - \underline{q}^{SB}) - u'(\underline{\theta} - \underline{q}^{SB})\right).$$

We saw above that it is impossible if $w(\bar{\theta}) = u(\bar{\theta} - \underline{\theta})$ and that it can become possible if $w(\bar{\theta}) = (1-p)u(\bar{\theta} - \underline{\theta})$ for p large enough.

An alternative formulation is

$$w(\theta) = (1-p)u(\theta) \qquad \text{for all } \theta.$$

Then, the rent of a good type verifies

$$u(\underline{\theta}) + u(\bar{\theta} - \underline{q}^{SB}) - u(\underline{\theta} - \underline{q}^{SB}) \geq u(\bar{\theta}) \geq (1-p)u(\bar{\theta}),$$

since $u(\bar{\theta} - q) - u(\underline{\theta} - q)$ is increasing in q. We always have Regime 0.

It is therefore important to understand how the imperfection of enforcement structures the outside opportunity levels of the reneging agents. We give below a few examples:

i) Black market.
Suppose that, when they carry out their contract, the principal and the agent can sell their share of the product on a idiosyncratic market each with the inverse demand function $P(\cdot)$ with $P(q)q$ concave.

Then

$$\begin{aligned} u(q) &= P(q)q \\ u(\theta - q) &= P(\theta - q)(\theta - q). \end{aligned}$$

Alternatively the agent may renege the contract and sell on a black market with the inverse demand function $\tilde{P}(\cdot)$. This provides an alternative revenue of $\tilde{P}(\theta)\theta$ or $(1-p)\tilde{P}(\theta)\theta$ if the agent is caught with probability p.

Again, Regime 0 cannot hold if

$$\tilde{P}(\underline{\theta})\underline{\theta} + P(\bar{\theta} - \underline{q}^{SB})(\bar{\theta} - \underline{q}^{SB}) - P(\underline{\theta} - \underline{q}^{SB})(\underline{\theta} - \underline{q}^{SB}) \leq \tilde{P}(\bar{\theta})\bar{\theta}.$$

In words it means simply that the information rent

$$P(\bar{\theta} - \underline{q}^{SB})(\bar{\theta} - \underline{q}^{SB}) - P(\underline{\theta} - \underline{q}^{SB})(\underline{\theta} - \underline{q}^{SB})$$

must be less than the difference of opportunities in the black market between the two types of agents. Clearly, all cases are possible depending on the demand function in the black market.

ii) Corruption of enforcer.
Suppose that leaving the country is only possible through corruption of the customs officer; and suppose that a minimal bribe of k in units of the good is needed to corrupt the officer. Assume $\underline{\theta} < k$. Then

$$\begin{aligned} w(\underline{\theta}) &= 0 \\ w(\bar{\theta}) &= u(\bar{\theta} - k). \end{aligned}$$

This yields a model analogous to the one studied above. Clearly, the details of the corruption game are crucial. For example, the agent might need the complicity of the officer to value the good and would have to share the good. Let α be the share of the agent, so that

$$w(\theta) = u(\alpha\theta).$$

Then for α large enough, Regime 0 cannot hold while it does if α is small enough.

iii) Joint Venture.
Suppose that to value his discovery the agent must pay a fixed cost k, that his principal and maybe others have already sunk.

Then $w(\theta) = u(\theta) - k$. The principal can implement Regime 0 and the more cheaply the larger is k. However, if other "principals" have also sunk the fixed cost, the agent may still renege and organize an auction between the principals. This would be the case of a researcher paid in his university to do research and who, after a large $\bar{\theta}$ discovery, would take advantage of the "no slavery" constraint in labor contracts to leave and offer his discovery to other universities. This illustrates the kind of inefficiency which may arise due to the "no slavery" condition.[8]

6.2 Continuous Case

The analysis can be easily extended to the case of continuous type θ in $[\underline{\theta}, \bar{\theta}]$. The objective function of the agent is then

$$U(\theta) = u(\theta - q(\theta)) + t(\theta)$$

resulting in an incentive constraint (for Regime 0)

$$\dot{U}(\theta) = u'(\theta - q(\theta)).$$

It is then easy to see that Regime 0 is impossible if the slope of the information rent $u'(\theta - q(\theta))$ is lower than the slope of the outside opportunity $w'(\theta)$.

For example, if $w'(\theta) = (1-p)u'(\theta)$

$$u'(\theta - q(\theta)) > (1-p)u'(\theta)$$

and Regime 0 always occur. If $w'(\theta) = u'(\theta - \underline{\theta})$ then for θ small,

$$u'(\theta - \bar{q}(\theta)) < u'(\theta - \underline{\theta})$$

as $q(\theta) < \theta$ for θ in a neighborhood of $\underline{\theta}$. Then Regime 0 is impossible for θ close to $\underline{\theta}$. For large θ it becomes possible (see Appendix C).

So what matters are not the absolute levels of information rents and outside opportunities, but their rate of growth in the parameter of asymmetric information.

7. CONCLUSION

We have studied a delegation problem in which the nonverifiability of the agent's discovery fuels the opportunistic behavior of agents who have high performances and may renege on their contract and value their discovery outside the principal-agent relationship. A striking implication of the optimal contract is that it may destroy the incentive of the principal to provide good working conditions to the agent which would increase the probability of high discovery, and even favor the reward for low discovery to discourage agents to exert high levels of effort which would increase this probability of high discovery. Then, we have shown how an improvement of institutions in the form of better enforcement of contract, brought about by computer equipment for example, may reverse those perverse incentives. Finally, we have shown the need for a deeper analysis of the transaction costs of contract enforcement whose details affect considerably the structure of the optimal contract and the incentives it creates.

Beyond this main point, our analysis calls for further research in various directions. One is in labor economics and R&D research where the traditional "no slavery" conditions which make easy for workers to end their employment relationship may have spectacular implications on incentives.

Another is the analysis of the impact of black markets on the structure of labor contracts in the formal economy. Also, it would be interesting to characterize the optimal auctions of contracts for discovery of resources when the nonverifiability and enforcement conditions of this paper hold.

Appendix A

The problem may be solved directly. Define

$$\varphi(q) = u(\bar{\theta} - q) - u(\underline{\theta} - q)$$

which is positive and increasing due to the concavity of u. The constraints may be rewritten

$$\underline{U} \geq 0 \qquad \bar{U} \geq w \qquad \varphi(\bar{q}) \geq \bar{U} - \underline{U} \geq \varphi(\underline{q})$$

Because the principal tries to reduce the rents, we get

$$\bar{U} = \max(w, \underline{U} + \varphi(\underline{q}))$$

and the constraints reduce to

$$\underline{U} \geq 0 \qquad \underline{U} \geq w - \varphi(\underline{q}) \qquad \varphi(\bar{q}) \geq \varphi(\underline{q})$$

from which we get

$$\underline{U} = \max(0, w - \varphi(\bar{q}))$$

and finally

$$\bar{U} = \max(w, \varphi(\underline{q})).$$

Hence the program reduces to

$$\nu\left[u(\bar{q})+u(\bar{\theta}-\bar{q})-\tfrac{1-\nu}{\nu}\max(0,w-\varphi(\bar{q}))\right]$$
$$+(1-\nu)\left[u(\underline{q})+u(\underline{\theta}-\underline{q})-\tfrac{\nu}{1-\nu}\max(w,\varphi(\underline{q}))\right]$$

to be maximized under $\bar{q} \geq \underline{q}$. In fact, the constraint can be ignored because maximization of each bracket yields $\bar{q} \geq q^*(\bar{\theta})$ and $\underline{q} \leq q^*(\underline{\theta})$. Notice that each bracket is concave in the example $u(x) = x - x^2/2$, because then φ is linear.

In the particular case $w = u(\Delta\theta)$, it is easily seen that the second bracket is maximized at $\underline{q} = q^*(\underline{\theta})$, which indeed verifies $\varphi(\underline{q}) \leq w$. A similar result holds for the first bracket if $\Delta\theta \geq \underline{\theta}$ (Regime 1). Otherwise one can solve

$$\max_q \; u(\bar{q})+u(\bar{\theta}-\bar{q})-\frac{1-\nu}{\nu}(w-\varphi(\bar{q}))$$

whose solution $\bar{q}_3$ may be such that $\varphi(\bar{q}_3) \leq w$ (Regime 3), or $\varphi(\bar{q}_3) > w$, and then we have Regime 2.

Appendix B

We can proceed as in Appendix A, taking into account the additional constraint

$$\bar{U} - \underline{U} \leq \frac{\psi}{\Delta\nu}.$$

We still have

$$\bar{U} = \max(w, \underline{U} + \varphi(\underline{q}))$$

and the constraints reduce to

$$\underline{U} \geq 0 \qquad \underline{U} \geq w - \varphi(\underline{q}) \qquad \varphi(\bar{q}) \geq \varphi(\underline{q})$$
$$\underline{U} \geq w - \frac{\psi}{\Delta\nu} \qquad \varphi(\underline{q}) \leq \frac{\psi}{\Delta\nu}$$

from which we get

$$\underline{U} = \max(0, w - \varphi(\bar{q}), w - \frac{\psi}{\Delta\nu})$$

and finally

$$\bar{U} = \max(w, \varphi(\underline{q})).$$

Hence the program reduces to

$$\nu\left[u(\bar{q})+u(\bar{\theta}-\bar{q})-\frac{1-\nu}{\nu}\max(0,w-\varphi(\bar{q}),w-\varphi(\underline{q})\leq\frac{\psi}{\Delta\nu})\right]$$

$$+(1-\nu)\left[u(\underline{q})+u(\underline{\theta}-\underline{q})-\frac{\nu}{1-\nu}\max(w,\varphi(\underline{q}))\right]$$

to be maximized under $\bar{q}\geq\underline{q}$ and $\varphi(\underline{q})\leq\frac{\psi}{\Delta\nu}$. Once more, the first constraint can be ignored. Compared to Appendix A, the new term in the first bracket tends to reduce $\bar{q}$; the second constraint also reduces $\underline{q}$, compared to the solution in Appendix A.

Appendix C: The continuous type case

Similar insights can be gained in the case where θ belongs to an interval $[\underline{\theta},\bar{\theta}]$ and is distributed according to the distribution $F(\theta)$ with a positive density $f(\cdot)$. We also assume the hazard rate properties

$$\frac{d}{d\theta}\frac{F(\theta)}{f(\theta)}\geq 0 \quad \text{and} \quad \frac{d}{d\theta}\frac{1-F(\theta)}{f(\theta)}\leq 0$$

to avoid bunching of a classical type.

It is easily shown that with an enforcement constraint of the type

$$U(\theta)\geq(1-p)u(\theta) \tag{C. 1}$$

we have no countervailing incentives.

PROPOSITION 4: *With the enforcement constraint (C. 1) the optimal contract entails downward distortion characterized by:*

$$u'(q^*(\theta))=u'(\theta-q^{SB}(\theta))-\frac{1-F(\theta)}{f(\theta)}u''(\theta-q^{SB}(\theta))$$

and a rent

$$U=(1-p)u(\underline{\theta})+\int_{\underline{\theta}}^{\theta}u'(\tau-q^{SB}(\tau))d\tau.$$

Enforcement constraints simply oblige the principal to give up an additional rent $(1-p)u(\underline{\theta})$ to each type.

With the enforcement constraint

$$U(\theta)\geq u(\theta-\underline{\theta}) \tag{C. 2}$$

one must distinguish two cases.

For $\Delta\theta$ small (more precisely for $\Delta\theta$ such that the solution $q(\theta)$ of the equation below is such that $q(\theta) < \underline{\theta}$) we have countervailing incentives.

PROPOSITION 5: *Under the enforcement constraint (C. 2), we have:*

- *For θ in $[\underline{\theta}, \theta_0]$, production $q_3(\theta)$ is upward distorted.*
- *For θ in $[\theta_0, \theta_1]$, there is bunching and $q_2(\theta) = \underline{\theta}$.*
- *For θ in $[\theta_1, \bar{\theta}]$, production $q_1(\theta)$ is downward distorted.*

Rents are

$$\begin{aligned} U(\theta) &= \int_{\underline{\theta}}^{\theta} u'(\tau - q_3(\tau))d\tau \textit{ for } \theta \textit{ in } [\underline{\theta}, \theta_0] \\ &= u(\theta - \underline{\theta}) \textit{ for } \theta \textit{ in } [\theta_0, \theta_1] \\ &= u(\bar{\theta}) - \int_{\theta}^{\bar{\theta}} u'(\tau - q(\tau))d\tau \textit{ for } \theta \textit{ in } [\theta_1, \bar{\theta}]. \end{aligned}$$

The main difference between the two types of enforcement constraints is as follows.

For (C. 1), the marginal utility of the resource is lower for the agent when for the principal. The solution entails simply a bonus for the agent which is paid each time there is a discovery whatever its value.

For (C. 2), the marginal utility of the resource is higher for the agent. Then, for discoveries of small variance, countervailing incentives prevail and lead to decrease the share of the resource left to the agent to increase his marginal utility for the good and decrease his rent.

For greater variance of discoveries, for which the marginal utility of the resource for the agent is higher and then lower than for the principal as θ increases, we have a complex solution sharing the features of the two cases above as θ increases with a bunching region in between.

Figure 5 describes the profile of quantities.[9]

$$\begin{aligned} u'(q_3(\theta)) &= u'(\theta - q_3(\theta)) + \frac{F(\theta)}{f(\theta)} u''(\theta - q_3(\theta)) \\ u'(q_0(\theta)) &= u'(\theta - q_0(\theta)) - \frac{1 - F(\theta)}{f(\theta)} u''(\theta - q_0(\theta)). \end{aligned}$$

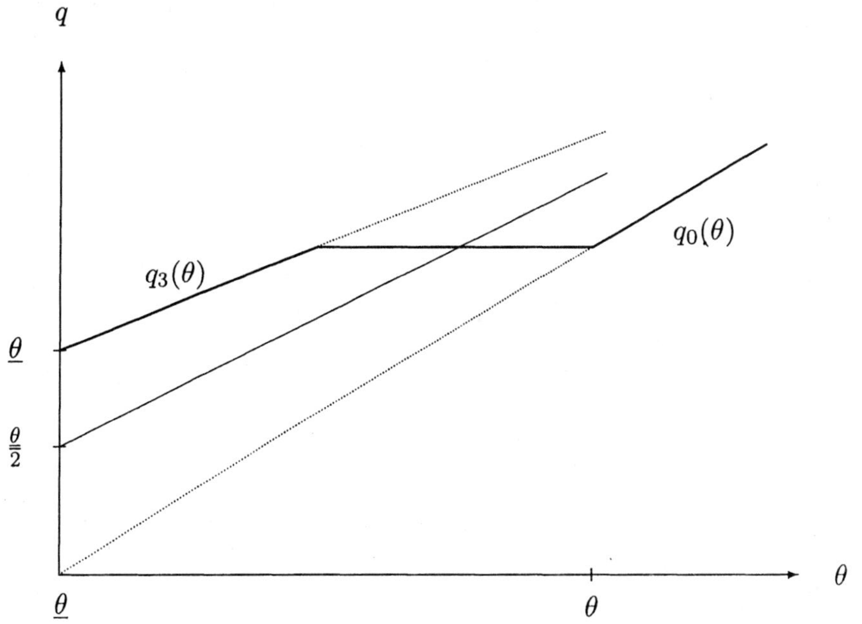

Figure 5. Profile of quantities in a continuous type case

Notes

1. The principal could avoid this ex post opportunistic behavior by requiring a bond which would be lost if reneging occurred. However the lack of wealth of the agent makes this strategy impossible.

2. We assume parameter values such that it is never in the principal's interest to opt for the shutdown of type $\underline{\theta}$ or for giving up the enforcement of type $\bar{\theta}$'s contract.

3. See for example Laffont and Martimort (2002), Chapter 2.

4. See Lewis and Sappington (1989), Maggi and Rodriguez (1995), Jullien (2000), and Laffont and Martimort (2002), Chapter 3 for a simple exposition.

5. Indeed define $g(q) = u(q) + u(\underline{\theta} - q)$. We have $\bar{V} - \underline{V} = g(\bar{q}) - g(\underline{q})$. Notice that $g(q)$ is decreasing for $q \geq \underline{\theta}/2$. The conclusion follows from $\bar{q} > \underline{q} = \underline{\theta}/2$ in regime 2 and 3.

6. Higher penalties are not possible because of limited liability constraints.

7. When the message space available to an agent depends on his type, the revelation principle may not hold (Green and Laffont, 1986). However, our simple case satisfies the necessary and sufficient condition obtained in Green and Laffont (1986) for the revelation principle to hold (see also Bull and Watson, 2001). The literature on manipulation of endowments (Postlewaite, 1979; Hurwicz et al., 1982) makes a crucial use of the inability to lie upward.

8. As labor contracts do not allow for such bonds, firms or universities often circumvent partially this problem by complementing the salary with financial advantages (like mortgage loans) which result in high penalties if the employment relationship is broken.

9. Additional assumptions are needed to avoid bunching when $u''(\cdot)$ is not constant.

References

Bull J. and J. Watson, 2001. 'Evidence Disclosure and Verifiability', mimeo, University of California at San Diego.

Becker, G. 1968. 'Crime and Punishment: An Economic Approach', *Journal of Political Economy*, 76, 169-217.

Becker, G. and G. Stigler, 1974. 'Law Enforcement, Malfeasance, and Compensation of Enforcers', *Journal of Legal Studies*, 3, 1-19.

Gibbard, A., 1973. 'Manipulation of Voting Schemes: A General Result', *Econometrica*, 41, 587-601.

Green, J. and J.J. Laffont, 1977. 'Characterization of Strongly Individually Incentive Compatible Mechanisms for the Revelation of Preferences for Public Goods', *Econometrica*, 45, 427-438.

Green, J. and J.J. Laffont, 1986. 'Partially Verifiable Information and Mechanism Design', *Review of Economic Studies*, 53, 447-456.

Hurvicz, L. E. Maskin and A. Postlewaite, 1982. 'Feasible Implementation of Social Choice Correspondences by Nash Equilibria', mimeo.

Jullien, B., 2000. 'Participation Constraint in Adverse Selection Models', *Journal of Economic Theory*, 93, 1-47.

Laffont, J.J., 2001. 'Enforcement, Regulation and Development', to appear in *Journal of African Economics*.

Laffont, J.J. and D. Martimort, 2002. *The Theory of Incentives*, Princeton University Press, Princeton.

Laffont, J.J. and J. Tirole, 1990. 'Bypass and Creamskimming', *American Economic Review*, 80, 1042-1061.

Lewis, T. and D. Sappington, 1989. 'Countervailing Incentives in Agency Problems', *Journal of Economic Theory*, 49, 294-313.

Maggi, G. and A. Rodriguez-Clare, 1995. 'On Countervailing Incentives', *Journal of Economic Theory*, 66, 238-263.

Myerson, R., 1979. 'Incentive Compatibility and Bargaining', *Econometrica*, 47, 61-73.

Postlewaite, A., 1979. 'Manipulation via Endowments', *The Review of Economic Studies*, 66, 255-262.

Stigler, G., 1970. 'Optimum Enforcement of Laws', *Journal of Political Economy*, 78, 526-536.

Chapter 3

RISK AVERSION AND GAINS FROM WATER TRADING UNDER UNCERTAIN WATER AVAILABILITY[1]

Javier Calatrava Leyva
Universidad Politécnica de Cartagena, Spain

Alberto Garrido
Universidad Politécnica de Madrid, Spain

1. INTRODUCTION

Water markets have been proposed as an instrument for dealing with water scarcity and improving economic efficiency in mature water economies subject to low levels of water reliability and to considerable uncertainty regarding their water supply. The potential welfare gains from the reallocation of water resources through voluntary exchange have been shown to be substantial (Vaux and Howitt, 1984; Rosegrant and Binswanger, 1994; Easter et al., 1998). These benefits are greater when supplies are reduced by the occurrence of a drought, its economic impact being mitigated (Miller, 1996). Properly designed and monitored, water trading provides some flexibility to water management, and may mitigate the adverse economic effects of droughts.

Most empirical studies dealing with the analysis of the potential welfare gains from water trading have neglected the uncertain nature of water availability. Simulations of water markets have rarely taken into account that water variability may have an effect on market activity. According to Antle's "risk-efficiency" hypothesis (1983), risk can affect economic efficiency both from the technical (productivity) or allocative (input decisions) points of view. In this sense, uncertain water availability has an impact in the production decisions of both risk-averse and risk-neutral producers.

The costs imposed by risk on a producer derive not only from the disutility of profit variability, but also from the effect that variability has on expected profit. Just (1975) shows that risk can influence decisions of a risk-neutral producer. A sufficient condition is that the random variable(s) affect non-linearly the producer's objective function. When the relation between the source of variability and profit is non-linear, then there is a cost derived of such risk even for risk-neutral producers (Just, 1975; Chambers, 1983; Babcock and Shogren, 1995). As expected profits are affected by risk, the way risk enters the profit function affects the modeling results and the policy implications (Antle, 1983).

In irrigated agriculture, most decisions regarding cropping schedule and several field operations are taken when the farmer is not sure about the definitive amount of water available for irrigation. In many Mediterranean areas, agricultural water availability exhibits a high level of interannual variability, so farmers generally face a considerable level of uncertainty about their final water allotment. Such uncertainty can be characterized by a probability distribution of allotments. Market exchanges usually take place when water allotments are known, but are partly subordinated to the decisions taken under uncertainty.

In the present chapter we analyze theoretically and empirically the effect of risk aversion on market participants' ex-ante optimal decisions and on the ex-post gains-from-trade under uncertain water supply. Studies dealing with water exchanges under supply uncertainty focus on its effects on water supplied by farmers. We broaden this scope to account for both sellers' and buyers' trading behavior, which are modeled as a tactical response available for farmers to cope with water uncertainty.

Two models have been developed to carry out the empirical application. One is a discrete stochastic programming model that simulates the behavior of irrigators under uncertainty and the other is a spatial equilibrium model to compute market exchange and equilibrium. Water market price endogeneity is modeled by an iterative process, which permits the characterization of price uncertainty from the results obtained from the spatial equilibrium model. Utility-efficient techniques are used to account for risk aversion. The empirical application is performed simulating a local water market in a district of the Guadalquivir Valley, Southern Spain.

2. LITERATURE REVIEW

Few studies have addressed theoretically the issue of uncertain water availability and its effect on production decisions. Howitt and Taylor (1993) analyze the case of a risk-averse producer that maximizes her expected utility (which depends on profit) in a context of uncertain water supply. They show that, in the optimum, the value of marginal product of water exceeds the expected shadow price of the resource. This implies that water is used less intensively than under certainty. It can be shown that the difference is larger the greater risk aversion is (Calatrava, 2002). In other words, production decisions are equivalent to those under certainty for an allotment below its expected value. This implies that the opportunity cost of water under uncertainty is also greater or equal than under certainty. Howitt (1998) points out that in a water market setting, uncertainty regarding water supply can be incorporated in the analytical framework in terms of an "annual scarcity cost". Thus, it can be assumed that, even in a dry year, water could be purchased from any source at some price. As we will show later, water supply uncertainty translates into uncertainty in the market price for water.

The effect of uncertain input prices on production decisions has been studied by Turnovsky (1969), Batra and Ullah (1974) and Blair (1974). Their main conclusion is that, under uncertainty regarding input price, input use, and therefore output produced, is lower for a risk-averse producer than in absence of such uncertainty. Batra and Ullah (1974) also show that under decreasing absolute risk aversion, the effect of an increase in uncertainty (a mean-preserving spread of the probability distribution of input price) results in a decrease in input use. Similarly, an increasing variance-preserving spread is identical to an increase in input price in the absence of uncertainty.

Most studies dealing with water transfers in a context of water uncertainty analyze the effect of such uncertainty on farmer's willingness to accept a compensation for transferring water to non-agricultural uses. Examples are the papers by Taylor and Young (1995), Turner and Perry (1997), Keplinger et al. (1998), Willis and Whittlesey (1998) and Knapp et al. (2003), where farmers' supply for water is derived using farm risk programming models. Decisions are taken under uncertainty regarding water availability[2]. Their main conclusion is that if uncertainty is not considered in the modeling process water supply is overestimated. Under uncertain water availability water supplied by sellers is therefore reduced. However, none of them simulates a wholesale market for water with producers that act both as buyers or sellers. Only Turner and Perry (1997) explicitly include market sales as an available choice for farmers. They do not model water exchanges, but they consider the possibility of buying or selling water at an exogenous deterministic constant price as a tactical response.

Some authors have criticized the excessive importance given in the farm modeling literature to risk preferences, to the detriment of the analysis of tactical responses or the characterization of probability distributions (Hardaker et al., 1991; Hardaker, 2000; Pannell et al., 2000; and Lien and Hardaker, 2001). According to this view, farmers are not that much interested in avoiding the risks they face, something that is not always possible, as in foreseeing their effects and being able to respond tactically by modifying their initial decisions as uncertainty reveals (Marshall et al., 1997). This is independent of their attitudes toward risk. Because a risk-neutral farmer also bears a cost from resource variability, which results in profit variability (Babcock and Shogren, 1995), it is in her interest to find feasible strategies to reduce such cost. This type of response is modeled by including tactical responses in farm programming models. Modeling tactical responses has a greater effect on model output because such responses tend to occur in extreme situations, when the effect of variability in expected profit is much greater than the effect of risk. Production strategies are modified to a larger extent than if risk aversion alone is considered (Hardaker et al., 1991; Babcock and Shogren, 1995; Pannell et al., 2000).

Uncertain water availability affects production decisions at the beginning of the cropping season, some of which can be lately modified or adjusted. Production decisions take into account not only their outcome but also the possible tactical responses the producer can resort to as uncertainty unfolds. Examples of tactical responses to a surplus or deficit of water with respect to the expected water availability are changes in water applications to crops, crop abandonment, or purchasing or selling water in the market. In presence of water markets, farmers would change their perception of the water supply uncertainty, focusing as well in the uncertainty of the market price to define their final water utilization (Calatrava, 2002). To model such type of tactical response, Discrete Stochastic Programming allows us to simulate the sequential nature of productive decisions taken in a context of uncertainty regarding water availability (Turner and Perry, 1997).

Market equilibrium problems are usually solved using endogenous price models, such as those developed by Enke (1951), Samuelson (1952) and Takayama and Judge (1964), to solve the problem of equilibrium in spatially separated markets. Such type of models have been used to simulate water allocation and market exchanges by Flinn and Guise (1970), Vaux and Howitt (1984) and Booker and Young (1994).

3. MODELLING THE BEHAVIOUR OF A SPOT MARKET PARTICIPANT UNDER UNCERTAIN WATER SUPPLY

Under certain water supply, a water right-holder that participates in a competitive annual spot water market faces the following problem:

$$Max_w \qquad \pi_m(w) = \pi(w) + P_m(A-w) \tag{1}$$

where w is the amount of water used in the production process; A is the allotment the producer is entitled to; P_m is the exogenous market price for water; $\pi_m(w)$ is the total profit function for the producer; and $\pi(w)$ denotes the profits derived from producing using w. $\pi(w)$ is a restricted profit function, with a negative second derivative (Cornes, 1992), that can be defined as:

$$\pi(w)=\{max_z\, pq(w, z) - \mathbf{c}'\mathbf{z} \,/\, \forall w\} \tag{2}$$

where $\mathbf{z}$ is a vector of inputs other than water; p is output price; $q(w,\mathbf{z})$ is the production function; and $\mathbf{c}'$ is the input costs vector. Therefore, it is assumed that profit function, $\pi(w)$, only depends on the amount of water used, being the optimal allocation of the inputs $\mathbf{z}$ implicit in the amount of water used. The term, $P_m(A-w)$, represents the cost incurred in buying water or the revenue from selling water in the market. In the optimum, a water user equals marginal profit and market price for water. Participation of a producer in a water market is influenced by all sources of risk and uncertainty, which may deviate her decisions from such optimum.

We assume that water allotment is uncertain at the time that production decisions are taken. We also assume that there is no possibility to rectify them or to change the production strategy, which implies that production decisions are irreversible. A consequence of these assumptions is that a producer positions herself in the market either as a buyer or as a seller at the beginning of the period depending on her expected water allotment. Production decisions are therefore taken under uncertainty, but market exchanges take place when uncertainty is unfolded. Thus, the amount of water exchanged would directly depend on the initial resource use decisions, on the amount of water available and on the market price for water.

It is also assumed that the producer has complete information regarding the probability distribution of water allotment. In practice, any producer can build up expectations about her water allotment depending on past years' allotments or the stock levels in the reservoirs she is serviced from. That is, she would take decisions taking into account a conditional probability distribution of water allotment with a minor range than the absolute probability distribution of allotment.

As commented above, it can be assumed that variability of allotment is transformed in the variability of the opportunity cost of water (Howitt and Taylor, 1993; Howitt, 1998). In a competitive water market, the cost of using water for production is given by its market price (purchase cost for the buyer and opportunity cost for the seller). In such context, a change in allotment implies a change in market price only, if it affects all users.

In that sense, it is important to clarify some points regarding the effect of a change in allotment, A. The amount of water available for a user determines its internal shadow price for water, and therefore her willingness to pay to acquire more or be compensated to use less. An increase in water allotment (ΔA), ceteris paribus, does not imply a change in the optimal amount of water used by a buyer, but an increase in the amount sold in the market, resulting in a profit gain of $P_m \cdot \Delta A$, as shown by Dinar and Letey (1991) and Weinberg et al. (1995). Similarly, an increase in allotment would not imply a change in the amount of water used by a buyer, but would certainly reduce the amount to be bought in the market. The underlying assumption, that the market price remains fix, is plausible only if the allotment of the producer changes. In practice, it can be expected that a significant variation in water availability affects the equilibrium market price for water, and therefore its dual value and optimal level of water use.

For a risk-neutral producer the problem is given by:

$$Max_w \qquad E[\pi_m(w)] = E[\pi(w) + P_m(A-w)] \tag{3}$$

And the first order conditions for problem (3) results in:

$$\pi'(w) = E[P_m] \tag{4}$$

The existence of a market for water eliminates the direct effect of uncertainty regarding allotment A on the producer's decisions, as it transfers the uncertainty to the price of water that will be formed in the market place. The problem is then reduced to that in which input price is uncertain.

A risk averse producer maximizes her expected utility without being certain of her water allotment and/or the market price of water, which are random variables (probability distribution known), as:

$$Max_w \qquad E[U(\pi_m(w; A, P_m))] = E[U(\pi(w) - P_m(w-A))] \tag{5}$$

where P_m is market price for water; A is water allotment; $\pi(w)$ is a restricted profit function that depends on the amount of water used in the production process, such as (2); and $\pi_m(w;A,P_m)$ is the profit function in presence of a market, whose argument is the amount of water used. Optimality conditions for problem (5) are derived in the appendix for different cases of uncertainty regarding parameters A y P_m.

If we assume **uncertainty in water allotment (*A*) and a known water price** (appendix , case 1), the first order condition (expression (A14)) is identical to that obtained for the case of certainty (marginal profit with respect to water use equal water price). Such theoretical market environment would be equivalent to a California-type Water Bank, in which an agency sets the price of the exchanges[3]. Such type of scheme eliminates the cost of uncertainty regarding water availability derived from non-optimal production decisions, and the producer chooses the amount of water that equals marginal profit from water use and its market price. A certain and institutionally-fixed water price would likely lead to non-clearing exchanged amounts, as the 1991 Californian water bank illustrates.

The of **uncertainty in both allotment (*A*) and water price (P_m)** is equivalent to a spot water market. As shown in the appendix (case 2), optimality condition for this case is given by expression (A22) as:

$$\pi'(w) - 2(w-E(A))REDQ \cdot V(P_m) + 3(w-E(A))^2 MSQ \cdot M_3(P_m) = E(P_m) \qquad (6)$$

Optimal production decisions differ from those under certainty. The difference depends on several factors: 1) the producer's utility function, characterized by its Risk Evaluation Differential Quotient (*REDQ*) and its Marginal Skewness Quotient (*MSQ*); 2) the level of risk in the random variable market price, P_m, characterized by its mean, variance and asymmetry; and 3) the positioning of the producer as a potential water buyer or seller, characterized by $w-E(A)$. As *REDQ* is positive and *MSQ* is negative for a risk averse producer, in the optimum, marginal profit from water use will be greater, equal or less than the expected market price for water depending on the values of the different components of expression (6).

When **$w-E(A)$ is positive**, that is, when the producer takes production decisions that imply using more water than her expected allotment, she positions herself as a potential water buyer. In this case, marginal profit $\pi'(w)$ will be greater than the expected water price $E[P_m]$. The producer will use less water than she would under certainty. The effect of uncertainty increases with larger used volumes above the expected allotment (that is, the amount of water she expects to buy), larger risk aversion, and greater variance and asymmetry of water prices. Only in the unlikely case of a extremely negative asymmetry of water price (probability of extremely low water prices) a potential buyer would use more water than under certainty.

On the other hand, when **$w-E(A)$ is negative**, that is, when the producer takes production decisions that imply using less water than her expected allotment, she positions herself as a potential water seller. In this case, marginal profit $\pi'(w)$ will be less than the expected water price $E[P_m]$. The producer will use more water than under certainty. The effect of uncertainty will be greater the smaller the amount of water used below the expected allotment is (that is, the less water she expects to sell), the more risk averse the producer is, and the greater the variance and asymmetry of water price are. Only in the unlikely case of an extremely negative asymmetry of water price, a potential seller would use more water than under certainty.

These conditions imply that both the amount of water to be supplied by sellers and to be demanded by buyers would be reduced as a consequence of uncertainty. Such reduction would be greater with greater market exposure (i.e. larger selling or buying positions). It would also augment with more risk aversion and more supply uncertainty. Results obtained for the case of a buyer are consistent with those of Batra and Ullah (1974) and Howitt and Taylor (1993). The result in the case of a seller is also similar to that of Turner and Perry (1997), taking into account that they only empirically analyse the effect of water uncertainty in risk-neutral farm programming models, resulting in an overestimation of farmers' water supply. If buyers use less water than in absence of uncertainty, the effect is a shift to the left of water demand. If sellers use more than in absence of uncertainty, then the effect of uncertainty is a shift to the left of water supply. As a result, market activity would be reduced. The effect on price is undetermined.

There is logic behind these results. When a producer is allowed to buy or sell water in a market, uncertainty in the amount of water she is entitled to does not influence directly the production decisions. It is uncertainty in water price and the buying or selling position which determine the optimal (expected utility maximization). This also explains the result obtained in the Water Bank case, where water price is previously known.

In figure 1, m is the difference between the amount of water used and water allotment ($m=w-A$), that is the excess demand and supply function. VMP_C is the inverse water demand for a producer in absence of uncertainty and VMP_U is the inverse water demand function for a producer under uncertainty. Depending on whether m is positive or negative, the producer acts as a buyer or as a seller in the market. As a result of uncertainty, both individual water demand and supply schedules become more inelastic. Figure 1 also shows that the difference between the marginal profit of water under uncertainty and under certainty is greater the more accentuated the buying or selling position is (i.e., the greater the absolute value of m is).

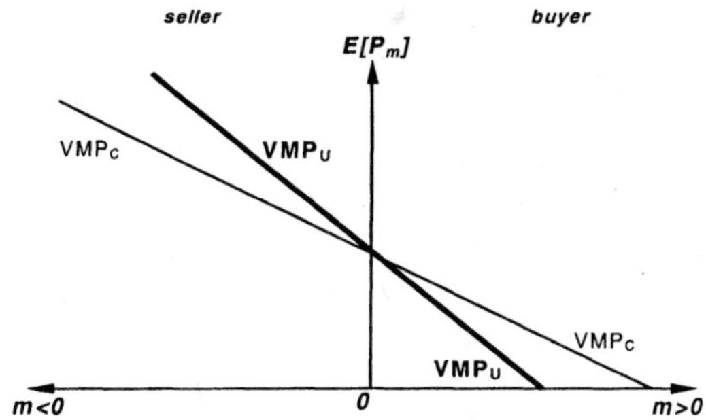

Figure 1. Excess water demand and supply function under uncertainty

If **uncertainty in water price (P_m) and a known water allotment** (A) is assumed (appendix, case 3), the optimal conditions obtained are given by expression (A30) as:

$$\pi'(w) - 2(w\text{-}A)REDQ\ V(P_m) + 3(w\text{-}A)^2 MSQ\ M_3(P_m) = E(P_m) \tag{7}$$

Expression (7) is similar to expression (6). The difference is that, as water allotment is known, the buying or selling position is denoted by (w-A) instead of (w-$E(A)$). The interpretation of (7) is similar to that of (6).

4. EMPIRICAL STUDY

An empirical application has been performed simulating a local water market among farmers belonging to the Guadalmellato irrigation district in the Guadalquivir River Basin (Southern Spain). Eleven different types of irrigated farms have been identified in the area. Their main characteristics are shown on Table 1.

Table 1. Farm types considered.

Farm type	Crops	Irrigation	Size (ha)	% in number	% of surface
1	Arable	Furrow	2	32.7	6.7
2	Arable	Sprinkler	2	25	5.2
3	Arable	Furrow	7	11.9	8.5
4	Arable	Sprinkler	7	6	4.3
5	Arable	Furrow	21	18.1	39
6	Arable	Sprinkler	21	2	4.2
7	Arable	Furrow	73	1	7.1
8	Arable	Sprinkler	73	0.3	2.4
9	Arable	Furrow	188	0.8	15.3
10	Citrus	Drip	52	0.8	4.3
11	Olive tree	Drip	20	1.5	3

Note: a total of 630 farms and 6129 hectares are considered.

Table 2. Scenarios of water uncertainty within a season

Scenario 1 (Stock < 40 Hm^3)		Scenario 2 (Stock > 40 Hm^3)	
Allotment (m^3/ha)	**Probability**	**Allotment (m^3/ha)**	**Probability**
4,900	0.250	5,500	0.4445
4,500	0.125	5,200	0.2222
4,000	0.125	4,500	0.2222
3,000	0.250	3,100	0.1111
500	0.125		
0	0.125		

Water uncertainty has been characterized using records from the basin authority regarding water stocks and releases of the "Guadalmellato" dam, from which the district is served, and individual water allotments for the period 1978-2000. At stage 1 (beginning of autumn), the level of water stored in the dam is in its lowest yearly levels. The basin authority identifies two possible scenarios. First, when reservoir's level is above 40 Hm^3, water releases for farmers when the irrigation season starts (in spring) can be determined with a high level of confidence. Second, for stock levels below 40 Hm^3 water supply reliability is very low, as the final release for irrigation depends entirely on winter rains which are subject to high variability. The series of farmers' annual allotments (in m^3 per ha) have been grouped within these two scenarios, and discrete probabilities have been calculated for each one. Table 2 shows these probabilities for each uncertainty scenario. For example, when the dam stock level at the beginning of autumn is less than 40 Hm^3, there is a 0.25 probability of the farmers receiving an allotment equal to 3,000 m^3/ha.

In the area of study, annual water allotment for farmers is usually determined in spring, after the autumn and winter rains. Uncertainty about water availability therefore directly affects production decisions regarding those crops that are planted on autumn (winter crops), and indirectly affects crops planted on spring through crop substitution effects. To account for this uncertainty and its effect of crop scheduling decisions, two different stages are considered (figure 2). On a first stage, cropping patterns are scheduled under uncertainty assuming perfect information with respect to the probability distribution of both water availability and water prices. Water buying or selling activities are therefore included in the model as tactical responses available for farmers. At a second stage, farmers may modify to a certain extent those initial decisions, once the definitive allotment is known. Choice variables are acreage and water applications to each crop in each stage, as well as water sold or bought in the market. Surface devoted to winter crops at the first stage cannot be modified, while the planned acreage devoted to spring and summer crops can be modified at the second stage. The analysis here is centered on uncertainty about water availability, the main risk-related issue of interest for farmers in the area, so other sources of risk are not considered.

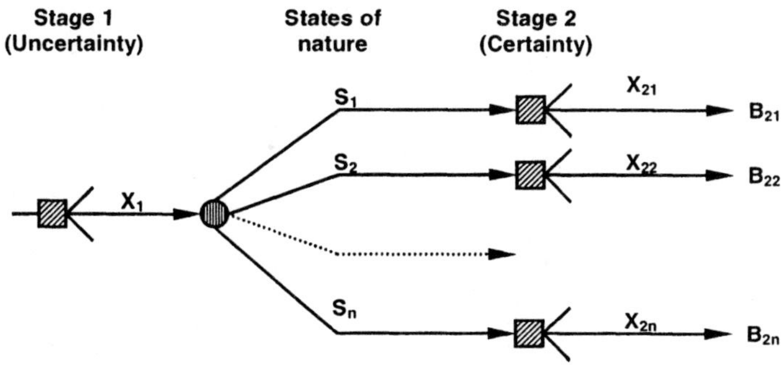

Figure 2. Decision tree for the uncertainty problem

Farmers' behavior under uncertain water supply and market water price is simulated using a **stochastic programming with recourse model** (SPR) with an **utility-efficient programming** (UEP) approach to account for risk attitudes (Patten et al., 1988). In the UEP approach, a SPR farm model is solved repeatedly assuming different values of the relative risk aversion coefficient. Model formulations are shown in table 3. The outcome from the stochastic model, solved for different risk aversion values, is compared with results from a deterministic model (CERT, denoting certainty) in which allocation decisions are taken under certainty. Both models provide profit and water demand functions that are used to simulate market exchanges.

The chosen utility function is a special form of the power utility function following recommendations by Lien and Hardaker (2001):

$$U(z,a) = \left(\frac{1}{1-\alpha}\right) z^{(1-\alpha)} \tag{8}$$

where z is annual net income and α is the coefficient of relative risk aversion. This utility function has a positive first derivative ($U'(z)>0$) and a negative second derivative ($U''(z)<0$), and presents decreasing absolute risk aversion (DARA) and constant relative risk aversion (CRRA).

The coefficient of absolute risk aversion for this utility function is:

$$r_a(z) = - U''(z) / U'(z) = \alpha / z \tag{9}$$

while the coefficient of relative risk aversion is:

$$r_r(z) = z\, r_a(z) = \alpha \tag{10}$$

Table 3. Models formulation.

	UEP model	CERT model
Objective function	$Max\Sigma_s Prob_s\ U(\pi_{is};r)$	$Max\pi_i$
	Constraints	
Profit definition	$\pi_{is}=\Sigma_k S_{2iks}[F_k(W_{2iks})P_k+EU_k-VC_{ik}(S_{2iks},W_{2iks})]$ $-FC_i-PM_sM_{2is} \quad \forall s$	$\pi_i=\Sigma_k S_{ik}[F_k(W_{ik})P_k+EU_k-VC_{ik}(S_{ik},W_{ik})]-FC_i$
Land availability	$\Sigma_k S_{tiks} \leq SAU_i \quad \forall t,s$	$\Sigma_k S_{ik} \leq SAU_i$
Water availability	$\Sigma_k S_{2iks}W_{2iks}-M_{2is}\leq SAU_iA_s \quad \forall s$	$\Sigma_k S_{ik}W_{ik} \leq SAU_iA_s$
Recursivity	$S_{1ik} = S_{2iks} \quad \forall k\in$ "winter crops" $,i,\ s$	
Non-negativity	S_{tiks} , $W_{tiks} \geq 0 \quad \forall i,t,k,s$	S_{ik} , $W_{ik} \geq 0 \quad \forall i,k$
Others	Agricultural policy	Agricultural policy

Decision variables	
S_{tiks}	area assigned on stage *t* by farm *i* to crop *k* under state of nature *s*
W_{tiks}	water assigned on stage *t* by farm *i* to crop *k* under state of nature *s*
M_{2is}	water bought or sold in the market in stage 2 by farm *i* under state of nature *s*
Parameters and functions	
$Prob_s$	Probability of state of nature *s*
R	Risk aversion coefficient
$F_k(W_{ik})$	Crop-water response function for crop *k*
P_k	Price for crop *k*
EU_k	Per hectare EU payment for crop *k*
$VC_{ik}(S_{ik},W_{ik})$	Variable costs for crop *k* and farm *i* (irrigation, harvest and transport included)
FC_i	Fixed costs for farm *i*
PM_s	Market price for water under state of nature *s*
SAU_i	Total area of farm *i*
A_s	Water allotment per hectare under state of nature *s*

Note: Winter crops are durum wheat, soft wheat, sugar beet, potato and garlic. The other crops are cotton, corn, sunflower, citrus and olive tree.

For a coefficient of relative risk aversion equal to 1 ($\alpha=1$) the utility function is the natural logarithm of profit ($U=ln\ z$). For risk neutrality the coefficient of relative risk aversion is zero ($\alpha=0$) and utility is equal to profit ($U=z$). The values considered for the coefficient of relative risk aversion with respect to wealth are in the range 0.5 (little risk aversion) to 4 (extremely risk averse), following recommendations by Hardaker et al (1997) and Hardaker (2000).

When uncertainty relates to income or profit, as is the case in this paper, Hardaker (2000) obtains the following expression (11), where W is wealth:

$$r_r(z) = z\ r_a(z) = z\ r_a(W) =(z/W)\ r_r(W) \tag{11}$$

Expression (11) is used to compute risk aversion coefficients $r_r(z)$. Wealth data for each farm is calculated using per hectare data for farms in the area of study from the Spanish Ministry of Agriculture (MAPA, monthly)[4].

Farm models have been calibrated to observed crop schedules for each farm type using Positive Mathematical Programming. Modeling has been performed for both scenarios shown in table 2. Water demand functions for each farm are derived from the above models, and used to simulate water exchanges in the market using a spatial equilibrium model described below.

The water market is simulated using an **endogenous price model** that maximizes economic surplus derived from market participation by all users, and can be written as:

$$\text{Max} \sum_i \left[\int_0^{m_i^*} f_i(m_i)\, dm_i \right] \tag{12a}$$

$$\text{s.t:} \qquad \Sigma_i\, m_i \leq 0 \tag{12b}$$

$$-m_i \leq A_i \ \forall i \tag{12c}$$

where $f_i(m_i)$ is the inverse demand function for water for user i (marginal profit); $m_i = w_i - A_i$ is the amount of water bought ($m_i > 0$) or sold ($m_i < 0$) in the market by user i; w_i is the total amount of water used by user i. The first constraint requires that all supplied water volumes be greater than or equal to the amounts demanded. The second constraint impedes a user to sell more water than her allotment A_i. Market price for water is derived from the dual value of the first constraint.

The water market model (12) provides the optimal allocation of water for each level of water availability (A), that is the amount of water bought or sold by each farm (m_i), and the equilibrium price for water (P_m). Profit from water use is calculated from the previously estimated profit functions using the amount of water used ($SAU_i\, A + m_i$) as argument; revenue or cost from selling or buying water in the market is calculated as $-m_i P_m$. Summing up these two terms yields total profits for each farm.

Market price, PM_s, is an exogenous stochastic parameter in each agent's decision model, but is dependent on the size of the allotment A. It results from the interaction of all market participants, being endogenous to the whole modeling process. In order to properly characterize uncertainty with respect to the price of water and make it endogenous an iterative process is used assuming perfect information regarding the probability distributions. The process is commented below and described in figure 3. Further details can be found in Calatrava (2002) and Calatrava and Garrido (2005a, 2005b).

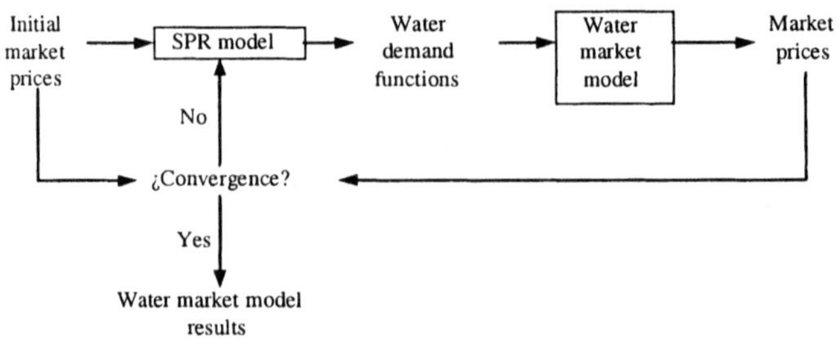

Figure 3. Iterative modeling process to characterize water price uncertainty

The procedure is launched considering some initial values of the market price for water under each state of nature. Then the farm model is run using those initial prices to characterize uncertainty regarding water price and water demand functions are obtained. Market exchanges are then simulated using those functions, and equilibrium prices obtained are used in the farm model to characterize uncertainty. This process is repeated until prices obtained from the market model converge with those used in the farm model from which demand functions were derived. The convergence criteria is that prices obtained for each state of nature differ in less than 0.001 euros/m^3 from those used to characterize uncertainty in the SPR model.

5. RESULTS AND DISCUSSION

Equilibrium market prices for water (reported in table 4) are similar for all models. UEP-r means that the SPR model has been solved assuming a "r" relative risk aversion coefficient. For example, in model UEP-4 a relative risk aversion coefficient equal to 4 has been assumed. Water prices diminish as the level of risk aversion increases under scarcity scenario 1. Prices for uncertainty scenario 2 are not shown as they barely differ for the different model assumptions. When uncertainty is higher, production decisions result in lower marginal profit of water and water is less valued, therefore reducing its scarcity price. Table 4 also shows the percentage of total water available that is exchanged in the market for uncertainty scenario 1. Risk aversion slightly reduces the level of market activity through its effect on water price.

Table 4. Percentage of total water available in the irrigation district exchanged in the market and water price for each level of water allotment (all assumptions). Scarcity scenario 1.

Allotment (10^3 m^3/ha)	Total water availability (hm^3)	CERT		UEP-1		UEP-4	
		% water exchanged	Price (€/m^3)	% water exchanged	Price (€/m^3)	% water exchanged	Price (€/m^3)
0	0	0	1.943	0.00	1.933	0.00	1.923
0.2	1,226	92.82	0.956	91.78	0.943	90.57	0.926
1	6,129	67.96	0.311	66.71	0.286	65.52	0.265
2	12,258	40.93	0.165	38.68	0.147	37.26	0.122
3	18,387	20.59	0.123	18.25	0.105	17.29	0.084
4	24,516	10.07	0.084	8.47	0.066	7.16	0.048
5	30,645	5.67	0.044	3.81	0.029	3.07	0.014
6	36,777	5.30	0.005				

We now look at the water market profit gains. Table 5 shows the levels of profit achieved through the water market for each model assumption and level of water allotment under uncertainty scenario 1. Profits are expressed as a percentage of the profit obtained from the deterministic model (CERT). Differences found among farms and scenarios deserve some comments.

Table 5. Water market profits for each farm type and model assumption (percentages of profit in absence of uncertainty). Supply uncertainty scenario 1.

Model	$r_r(W)$	Farm type 1	2	3	4	5	6	7	8	9	10	11
CERT		100	100	100	100	100	100	100	100	100	100	100
UEP-0	0	93.2	91.3	96	95.9	96.3	96.1	96.5	96.1	95.9	102	101
UEP-0.5	0.5	93	91.1	95.6	95.4	95.6	95.4	96.2	95.8	95.5	102.3	101.2
UEP-1	1	92.7	90.6	94.9	94.6	94.7	94.6	95.8	95.4	95	102.6	101.5
UEP-2	2	91.9	89.7	93.5	93.1	92.8	92.6	94.7	94.3	93.8	103.4	102
UEP-3	3	90.9	88.7	91.9	91.4	90.3	90	93.4	92.7	92.5	104.2	102.6
UEP-4	4	89.5	87.5	90.1	89.5	87	86.6	91.8	90.7	90.1	105.1	103.3

Under uncertainty scenario 1, profit is reduced with respect to that of model CERT. Farm profits achieved through the market are reduced under uncertainty. If risk neutrality is considered ($r_r(w)$=0) uncertainty causes a reduction in profits ranging from 3.5% (farm type 7) to 8.7% (farm type 2). Reductions in profits grow with more risk aversion. For an extreme level of risk aversion ($r_r(w)$=4) profits are reduced to levels ranging from 8.2% (farm type 7) to 13.4% (farm type 6). For intermediate levels of risk aversion ($r_r(w)$ between 1 and 2), profits are slightly below those for the risk-neutrality case, and most of the adverse effect of uncertainty would be independent of farmers' attitudes toward risk.

A noteworthy exception is found in the case of farm types 10 and 11 that present higher profit with than without uncertainty. The reason is that the stochastic decision model does not allow them to reallocate their land among crops as all their area is permanently devoted to citrus and olive trees respectively. The effect of uncertainty over their market profits is indirectly given by the effect of such uncertainty on the remaining nine types of farms and therefore on the market equilibrium. Uncertainty reduces market profit for the other nine farms that take less efficient production decisions than under certainty, and market price is reduced. As farm types 10 and 11 are always water buyers, they purchase it at lower prices and increase their market profit. These results show that risk not only has an impact on water market efficiency gains, but it also has distributive effects. Buyers benefit from sellers becoming less efficient because of increasing risk in water supply or risk aversion. A more general formulation of this conclusion is that users with inelastic water demands would become winners of an increasingly risk environment, at the cost of trading partners with more elastic demands.

Figures 4 show water market profits for both scenarios aggregated at the district level. Profits are expressed as a percentage of the profit from model CERT (a 100% level flat line). Results in these figures confirm those above.

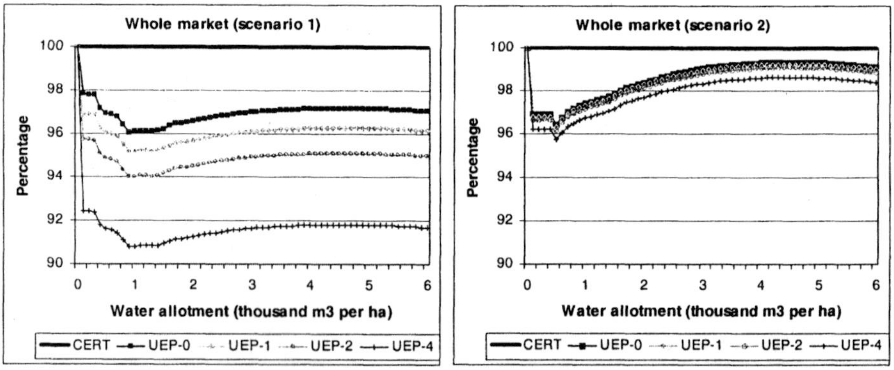

Figures 4. Water market profits for each assumption and uncertainty scenario (percentages of profit in absence of uncertainty, model CERT=100)

6. CONCLUSIONS

Uncertainty in water availability reduces farmers' benefits because they must take ex–ante decisions that may be irreversible. As water marketing decisions are derived from previous production decisions, uncertainty in water supply will result in a reduction of gains from trade. It has been shown analytically that the possibility of entering a spot water market eliminates the effect of the uncertainty directly derived from variable water availability, as it influences production decisions taken by a producer indirectly through its effect on market price for water. This implies that a Water Bank such as the one in California, where water price was certain, eliminates the effect of uncertainty in water availability for users, although the allocation of water would not be efficient, as demand and supply would not match.

For a risk neutral producer, the optimal production decisions imply that the marginal profit be equal to the expected market price for water. The cost of uncertainty for a risk neutral producer would be given by the variability of water availability (i.e. the level of risk exposure, analytically by the actual deviations of the market price for water from its mathematical expectations).

For a risk averse producer the cost of uncertainty would be also given by the disutility that it causes. Uncertainty and risk aversion result in production decisions that maximize expected utility for the producer. It has been shown that this additional cost derives from the producer's risk attitudes, from her level of buying or selling position in the water market, and from the levels of risk she is exposed to.

If ex-ante production decisions by a risk averse market participant are considered, expected utility is maximized for a water buyer (seller) when marginal profit derived of the productive water use is greater (lesser) than the expected market price for water $E[P_m]$. Under water supply and price uncertainty, a producer that plans to purchase water will use less water than she would under certainty for a price equal to $E[P_m]$. On the other hand, a producer that decides not to use her full expected allotment will use more water than she would under certainty for a price equal to $E[P_m]$. Therefore, both water demanded by buyers and water supplied by sellers are reduced and their demand and supply schedules become more inelastic. Once uncertainty is unfolded, farmers can modify to a certain extent their initial production decisions, to reduce water use and sell it, or to increase water use by entering the market as buyers. Nevertheless, water market profits will be reduced with respect to those under certainty, being smaller with greater uncertainty and risk aversion.

The empirical results show that water exchanged and water prices are slightly reduced as a result of uncertainty. This reduction is greater the greater uncertainty and risk aversion are. Reductions in farm profits achieved through the market are proportional to the level of risk in water availability and to risk attitudes. For average levels of risk aversion, profits are slightly below those for the risk-neutrality case, implying that most of the adverse effect of uncertainty is independent of farmers' attitudes towards risk.

Appendix. Expected utility maximization for a spot water market participant.

A utility function with just one argument, such as profit, can be expressed as a function of the moments of the probability distribution of such attribute using a Taylor series expansion around its mean value, as shown by Anderson et al. (1977, chapter 4):

$$U[\pi]=U[E(\pi)]+U_2[E(\pi)]M_2(\pi)/2+\ldots+U_n[E(\pi)]M_n(\pi)/n! \qquad \text{(A.1)}$$

where U is an utility function that depends on profit π; $M_k(\pi)$ is the kth order moment around the mean value of π; and U_k is the kth derivative of utility function U. As it can be seen in the above expression, the moments of the probability distribution of profit determine the utility level that the producer obtains from such profit. It is generally assumed that terms of an order greater than 3, or even 2, provide little accuracy to the approximation to the value of utility, being usually disregarded.

Using a parallel approach to that of Anderson et al. (1977, chapter 6) for the problem of expected utility maximization under product price and production risk, we will now derive first order conditions for the case of a producer that takes water trading decisions. Her problem is to maximize her expected utility from profit:

$$E[U(\pi_m(w;\ A))] = E[U(\pi(w) - P_m(w\text{-}A))] \qquad \text{(A.2)}$$

where $\pi_m(w;\ A)$ is total profit function; $\pi(w)$ is profit function derived from using the amount of water w for production; U is the utility function for a risk averse producer, utility that is concave and depends on profit; P_m is market price for water; y A is producer's water allotment. Although profit $\pi(w)$ depends on water used w, in the following analysis we will drop the argument to simplify the notation.

If we take the first three terms of the Taylor series expansion of utility function U, we can express utility as a generic function of mean profit, variance of profit and the asymmetry or 3rd order moment of profit as:

$$Utility \equiv U[E(\pi_m),\ V(\pi_m),\ M_3(\pi_m)] \qquad \text{(A.3)}$$

Maximizing expression (A.3) with respect to variable w:

$$dU/dw = [\partial U/\partial E(\pi_m)][dE(\pi_m)/dw] + [\partial U/\partial V(\pi_m)][dV(\pi_m)/dw] + [\partial U/\partial M_3(\pi_m)][dM_3(\pi_m)/dw] \equiv 0 \qquad \text{(A.4)}$$

Dividing (A.4) by $[\partial U/\partial E(\pi_m)]$:

$$dE(\pi_m)/dw + [\partial U/\partial V(\pi_m)]/[\partial U/\partial E(\pi_m)][dV(\pi_m)/dw] \\ +[\partial U/\partial M_3(\pi_m)]/[\partial U/\partial E(\pi_m)][dM_3(\pi_m)/dw] \equiv 0 \qquad (A.5)$$

Expression (A.5) can be written as:

$$dE(\pi_m)/dw - REDQ[dV(\pi_m)/dw] - MSQ[dM_3(\pi_m)/dw] \equiv 0 \qquad (A.6)$$

where:

- $REDQ=-[\partial U/\partial V(\pi)]/[\partial U/\partial E(\pi)]$ is called the "Risk Evaluation Differential Quotient", and measures the quotient among marginal utility of the variance of profit and marginal utility of mean profit. As marginal utility of profit is positive for a risk averse producer, while marginal utility of the variance of profit is negative, *REDQ* will be positive.
- $MSQ=-[\partial U/\partial M_3(\pi)]/[\partial U/\partial E(\pi)]$ is called the "Marginal Skewness Quotient", and measures the quotient between marginal utility of the 3rd-order moment of profit and marginal utility of expected profit. For a risk averse producer, the marginal utility of $M_3(\pi)$ will be positive. The more positive skewness of profit is, the probability of occurrence of lower levels of profit will get reduced, ceteris paribus. Therefore, if $\partial U/\partial M_3(\pi)>0$, then $MSQ<0$ for a risk averse producer.

Now we calculate first order condition (A.5) for the different cases of uncertainty in allotment and future market price for water.

Case 1: Market price for water is known, while allotment *A* is a random variable.

From expression (A.2) we calculate the mean value, the variance and the 3rd order moment of profit π_m respectively as:

$$E[\pi_m] = \pi(w) - P_m\,(w-E(A)) \qquad (A.7)$$
$$V[\pi_m] = V[P_m A] = P_m^2 V(A) \qquad (A.8)$$
$$M_3[\pi_m] = M_3[P_m A] = P_m^3 M_3(A) \qquad (A.9)$$

Taking derivatives in (A.7), (A.8) and (A.9) with respect to *w* :

$$dE(\pi_m)/dw = \pi'(w) - P_m \qquad (A.10)$$
$$dV(\pi_m)/dw = 0 \qquad (A.11)$$
$$dM_3(\pi_m)/dw = 0 \qquad (A.12)$$

Substituting (A.10), (A.11) and (A.12) in (A.6), and rearranging:

$$\pi'(w) = P_m \qquad \text{(A.13)}$$

Case 2: Both allotment *A* and water price are uncertain (it is assumed that both are stochastically independent random variables).

From expression (A.2) we calculate the mean value, the variance and the 3rd order moment of profit π_m respectively as:

$$E[\pi_m] = \pi(w) - E(P_m)(w\text{-}E(A)) = \pi(w) - E(P_m)w + E(P_m)E(A)) \qquad \text{(A.14)}$$

$$V[\pi_m]=V[\pi(w)\text{-}P_m(w\text{-}A)]=[E(P_m)]^2V(w\text{-}A)+[E(w\text{-}A)]^2V(P_m)+ +V(w\text{-}A)V(P_m)=[E(P_m)]^2V(A)+[E(w\text{-}A)]^2V(P_m)+V(A)V(P_m) \qquad \text{(A.15)}$$

$$M_3[\pi_m]=M_3[\pi(w)\text{-}P_m(w\text{-}A)]=M_3[\text{-}P_m(w\text{-}A)]= \text{-}M_3[P_m(w\text{-}A)]= =\text{-}[E(P_m)]^3 M_3(w\text{-}A) \text{-}[E(w\text{-}A)]^3 M_3(P_m) \text{-}M_3(w\text{-}A)M_3(P_m)= =[E(P_m)]^3 M_3(A) \text{-}[E(w\text{-}A)]^3 M_3(P_m)+ M_3(A)M_3(P_m) \qquad \text{(A.16)}$$

Taking derivatives in (A.14), (A.15) and (A.16) with respect to *w*:

$$dE(\pi_m)/dw = \pi'(w) - E(P_m) \qquad \text{(A.17)}$$

$$dV(\pi_m)/dw = 2(w\text{-}E(A))V(P_m) \qquad \text{(A.18)}$$

$$dM_3(\pi_m)/dw=\text{-}3(w\text{-}E(A))^2M_3(P_m) \qquad \text{(A.19)}$$

Substituting (A.17), (A.18) and (A.19) in (A.6), and rearranging:

$$\pi'(w) -2(w\text{-}E(A))REDQ\ V(P_m)+3(w\text{-}E(A))^2MSQ\ M_3(P_m) = E(P_m) \qquad \text{(A.20)}$$

Case 3: Water allotment *A* is known, while market price for water is a random variable.

From expression (A.2) we calculate the mean value, the variance and the 3rd order moment of profit π_m respectively as:

$$E[\pi_m] = \pi(w) - E(P_m)(w\text{-}A) = \pi(w) -E(P_m)w +E(P_m)A \qquad \text{(A.21)}$$

$$V[\pi_m]=V[\pi(w)\text{-}P_m(w\text{-}A)]= (w\text{-}A)^2V(P_m) \qquad \text{(A.22)}$$

$$M_3[\pi_m]=M_3[\pi(w)\text{-}P_m(w\text{-}A)] =M_3[\text{-}P_m(w\text{-}A)] = \text{-}(w\text{-}A)^3M_3(P_m) \qquad \text{(A.23)}$$

Taking derivatives in (A.21), (A.22) and (A.23) with respect to *w*:

$$dE(\pi_m)/dw = \pi'(w) - E(P_m) \qquad \text{(A.24)}$$

$$dV(\pi_m)/dw = 2(w\text{-}A)V(P_m) \qquad \text{(A.25)}$$

$$dM_3(\pi_m)/dw=\text{-}3(w\text{-}A)^2M_3(P_m) \qquad \text{(A.26)}$$

Substituting (A.24), (A.25) and (A.26) in (A.6), and rearranging:

$$\pi'(w) -2(w\text{-}A)REDQ\ V(P_m)+3(w\text{-}A)^2MSQ\ M_3(P_m) = E(P_m) \qquad \text{(A.27)}$$

Notes

1. This chapter expands previous work carried out by the authors and published in Calatrava (2002) and Calatrava and Garrido (2005a and 2005b).

2. Discrete Stochastic Programming (DSP) or Stochastic Programming with Recourse (SPR), developed by Cocks (1968) and Rae (1971), appears as the most adequate method to model uncertainty in resource availability and input prices. It has been used by Taylor and Young (1995), Turner and Perry (1997) and Keplinger et al. (1998) to model uncertain irrigation water availability.

3. In a water bank a different price is set for buyers and sellers, to cover transaction and transport costs. However, from the point of view of an individual producer's behavior such price difference does not influence the analysis.

4. The authors wish to thank Nabil Balti for collecting and compiling these data.

References

Anderson, J.R., J.L. Dillon, and B. Hardaker, 1977. *Agricultural decision analysis,* Iowa State University Press, Ames.

Antle, J.M., 1983. 'Incorporating risk in production analysis', *American Journal of Agricultural Economics*, 65(5), 1099-1106.

Babcock, B. A. J.F. and Shogren, 1995. 'The cost of agricultural production risk', *Agricultural Economics*, 12(2), 141-150.

Batra, R.N. and A. Ullah, 1974. 'Competitive firm and the theory of input demand under price uncertainty', *Journal of Political Economy*, 82 (3), 537-548.

Blair, R.D., 1974. 'Random input prices and the theory of the firm', *Economic Inquiry*, 2 (3), 214-225.

Booker, J.F. and R.A. Young, 1994. 'Modeling intrastate and interstate markets for Colorado river water resources', *Journal of Environmental Economics and Management*, 26(1), 66-87.

Calatrava, J., 2002. *Los mercados de agua en la agricultura y el riesgo económico: Una aplicación en el valle del Guadalquivir*, Doctoral dissertation, Department of Agricultural Economics, Universidad Politécnica de Madrid.

Calatrava, J. and A. Garrido, 2005a. 'Spot water markets and risk in water supply', *Agricultural Economics*, 33, 131-143.

Calatrava, J. and A. Garrido, 2005b. 'Modelling water markets under uncertain water supply', *European Review of Agricultural Economics*, 32, 119-142.

Cocks, K.D., 1968. 'Discrete stochastic programming', *Management Science*, 15, 72-79.

Cornes, R., 1992. *Duality and modern economics*, Cambridge University Press, Cambridge.

Chambers, R.G., 1983. 'Risk in agricultural production: Discussion', *American Journal of Agricultural Economics*, 65(5) , 1114-1115.

Dinar, A. and J. Letey, 1991. 'Agricultural water marketing, allocative efficiency and drainage reduction', *Journal of Environmental Economics and Management*, 20, 210-223.

Easter, K.W., M.W. Rosegrant and A. Dinar (eds.), 1998. *Markets for Water: Potential and Performance*, Kluwer Academic Publishers, Boston.

Enke, S., 1951. 'Equilibrium between spatially separate markets: solution by electric analogue', *Econometrica*, 19, 40-47.

Flinn, J.C. and J.W.B. Guise, 1970. 'An application of spatial equilibrium analysis to water resource allocation', *Water Resources Research*, 6(2), 398-409.

Hardaker, J.B., 2000. *Some issues in dealing with risk in agriculture*, Working Paper 2000-3, Graduate School of Agricultural and Resource Economics, University of New England, Armidale.

Hardaker, J.B., R.B.M. Huirne and J.R. Anderson, 1997. *Coping with risk in Agriculture*, CAB International, Wallingford.

Hardaker, J.B., S. Pandey and L.H. Patten, 1991. 'Farm planning under uncertainty: a review of alternative programming models', *Review of Marketing and Agricultural Economics*, 59(1) , 9-22.

Howitt, R., 1998. 'Spot prices, option prices, and water markets: an analysis of emerging markets in California', in *Markets for Water: Potential and performance*, edited by K. W. Easter, M. Rosegrant and A. Dinar, Kluwer Academic Publishers, New York.

Howitt, R. and C.R. Taylor, 1993. 'Some Microeconomics of Agricultural Resource Use' in *Agricultural and Environmental Resource Economics*, edited by G.A. Carlson, D. Zilberman and J.A. Miranowski, Oxford University Press, Oxford.

Just, R.E., 1975. 'Risk aversion under profit maximization', *American Journal of Agricultural Economics*, 57(2), 347-352.

Keplinger, K.O., B.A. McCarl, M.E. Chowdhury. and R.D. Lacewell, 1998. 'Economic and hydrologic implications of suspending irrigation in dry years', *Journal of Agricultural and Resource Economics*, 23(1) , 195-205.

Knapp, K.C., M. Weinberg, R.E. Howitt and J.F. Posnikoff, 2003. 'Water transfers, agriculture, and groundwater management: A dynamic economic analysis', *Journal of Environmental Management*, 67, 291-301.

Lien, G. and J.B. Hardaker, 2001. 'Whole-farm planning under uncertainty: impacts of subsidy scheme and utility function on portfolio choice in Norwegian agriculture', *European Review of Agricultural Economics*, 28(1), 17-36.

MAPA (monthly). *Boletín Mensual de Estadística Agraria.* Ministerio de Agricultura, Pesca y Alimentación, Madrid.

Marshall, G.R., R.E. Jones and L.M. Wall, 1997. 'Tactical opportunities, risk attitude and choice of farming strategy: an application of the distribution method', *Australian Journal of Agricultural and Resource Economics*, 41(4), 499-519.

Miller, K.A., 1996. 'Water Banking to Manage Supply Variability' in *Advances in the Economics of Environmental Resources, Vol. 1: Marginal Cost Rate Design and Wholesale Water Markets*, edited by D.C. Hall, JAI Press, Greenwich, 195-210.

Pannell, D.J., B. Malcolm and R.S. Kingwell, 2000. 'Are we risking too much? Perspectives in farm modelling', *Agricultural Economics*, 23(1), 69-78.

Patten, L.H., J.B. Hardaker and D.J. Pannell, 1988. 'Utility-efficient programming for whole-farm planning'. *Australian Journal of Agricultural Economics*, 32, 88-97.

Rae, A., 1971. 'Problem stochastic programming, utility, and sequential decision problems in farm management', *American Journal of Agricultural Economics*, 53(3), 448-460.

Rosegrant, M.W. and H.P. Binswanger, 1994. 'Markets in tradable water rights: Potential for efficiency gains in developing country water resource allocation', *World Development*, 22, 1613-1625.

Samuelson, P.A., 1952. 'Spatial price equilibrium and linear programming', *American Economic Review*, 42(2), 283-303.

Sandmo, A., 1971. 'On the theory of the competitive firm under price uncertainty', *American Economic Review*, 61(1), 65-73.

Takayama, T. and G.C. Judge, 1964. 'Spatial equilibrium and quadratic programming', *Journal of Farm Economics*, 46(1), 67-93.

Taylor, R. G. and R.A. Young, 1995. 'Rural-to-urban water transfers: measuring direct foregone benefits of irrigation water under uncertain water supplies', *Journal of Agricultural and Resource Economics*, 20(2), 247-262.

Turner, B. and G.M. Perry, 1997. 'Agricultural to instream water transfers under uncertain water availability: a case study of the Deschutes river, Oregon', *Journal of Agricultural and Resource Economics*, 22(2), 208-221.

Turnovsky, S.J., 1969. 'The behaviour of a competitive firm with uncertainty in factor markets', *New Zealand Economic Papers*, 3(1), 52-58.

Vaux, H.J. and R.E. Howitt, 1984. 'Managing water scarcity: An evaluation of interregional transfers', *Water Resources Research*, 20(7), 785-792.

Weinberg, M., C. L. Kling and J.E. Wilen, 1995. 'Water markets and water quality', *American Journal of Agricultural Economics*, 75(2), 278-291.

Willis, D.B. and N.K. Whittlesey, 1998. 'Water management policies for streamflow augmentation in an irrigated river basin', *Journal of Agricultural and Resource Economics*, 23(1), 170-190.

Chapter 4

DYNAMIC UNCERTAINTY AND THE PRICING OF NATURAL MONOPOLIES: THE CASE OF URBAN WATER MANAGEMENT

Ephraïm Clark
Middlesex University, UK

Gérard Mondello
CNRS, GREDEG, France

1. INTRODUCTION

Because of its specific features and structural organization, the local drinking water service works as a natural monopoly. Hence, the service is often managed and supplied by government: municipalities (or a union of municipalities) under the authority of the mayor and the municipal council in the developed world and nationalized public utilities in the developing world. However, the process of privatization and deregulation in the eighties (in the developed world) and nineties (in the developing world) has called into question the role of government as a service provider and made privatization the order of the day. The situation of privatization has been complicated, however, by the fact that water management has evolved to the point where it requires increasingly sophisticated knowledge and technologies related to a high level of very complex research and development that are protected by patents, thereby creating de facto monopolies in technology.

Privatization has also raised a number of sensitive issues when applied to water management. It is well known that water is a specific economic good that is indispensible to life itself. Thus, where water management and supply are concerned, there are moral and ethical concerns that go beyond the simple criteria of economic efficiency that should accompany privatization. Since water is an essential good, authorities should guarantee its provision. This involves laying down specific policies that guarantee the provision of water with respect to quantity and quality and allows its consumption for basic human needs at affordable prices. In the context of privatization, this obligation puts the government in a position where it must reconcile a number of conflicting requirements (political, environmental, social, strategic, ethical and even sometimes religious). The means of this conciliation pass through making the rules of management of the natural monopoly compatible with these requirements. In this paper, we analyze the privatization of water management with respect to the pricing of potable water in the context of its status as a natural monopoly and as indispensible to life.

To this end, we review the standard pricing approaches, Ramsey-Boiteux marginal cost pricing and non-linear pricing, and show how they can limit the negative effects of a pure monopoly. We also show their limitations, notably that they are static and relevant to only one state of nature. They also fail to deal with the important question of price that the private firm should pay for the right to exploit the monopoly. We then present a model based on standard techniques in real option theory that can be used as a pricing program that gives a fair deal to both the regulator and the delegated firm. Our pricing program goes beyond marginal cost and non-linear pricing by introducing time and risk. Importantly, it provides a solution to the price that the delegated firm should pay for the right to exploit the monopoly. It also includes the relatively recent problem of technological monopolies that make it possible for the private water firms to extract supplementary economic rents. The fair deal in our pricing program eliminates economic rents while rewarding the risk taken on by the firm. It also gives the value of the monopoly that the regulator should receive as payment from the firm for the right to exploit the monopoly. In this sense what we propose is a fair deal.

The rest of the paper is organized as follows. In section 2, we review the relevant aspects of potable water supply as a natural monopoly. Section 3 outlines the standard pricing paradigmes and analyzes the policy implications for water management. In section 4 we present the pricing paradigm using real options. Section 5 concludes the paper.

2. THE SUPPLY OF DRINKING WATER AS A NATURAL MONOPOLY

The distribution of drinking water obeys the theoretical rules of a natural monopoly. From an industrial and economic point of view, this does not mean that it is a straightforward exercise to determine the strategic variables such as the pricing system or the proper investment level that ensures the equity of distribution and the improvement of quality that maximizes social welfare. On the contrary, the unique status of water as the source of human life only complicates the problem. Therefore, in this section we seek to outline the analytical problems associated with natural monopolies as they apply to the supply of drinking water.

2.1 Water management as a natural monopoly

Because private companies, which obey rules dictated by the market, have become more active in water management, the distribution and treatment of drinking water are increasingly considered as industrial activities. This development coincides with the changing definition of the concept of a natural monopoly, which has evolved to take competition into account. The role of competition in the allocation of a natural monopoly as such, when the local authority seeks to avoid direct management, raises a whole series of contemporary questions. Thus, in a first step, we look at these questions and the modern definition of a natural monopoly in the context of the management of drinking water.

The modern definition of a natural monopoly describes production in three dimensions: cost, output and scale. To synthesize these concepts, Baumol et al. (1982) introduced the notion of cost sub-additivity. It specifies that, for a given production level and product mix, if an individual firm has lower production costs than the sum of its competitors, the individual firm is in a situation of natural monopoly.

Consider firm 1 with respect to n-1 competing firms indexed by $i = 2,\ldots,n$ that produce $q_1, q_i, \in R^m$, $i = 2,\ldots,n$, where m is the number of activities. Then, for $\overline{q}_1, \overline{q}_i, i = 2,\ldots,n$ such that $\overline{q}_1 = \sum_{i=2}^{n} \overline{q}_i$, firm 1 is a natural monopoly if, considering the associated cost structure where $C(\overline{q}_1)$, and $C(\overline{q}_i)$, $i = 2,\ldots,n$ are the cost functions of firm 1 and the n-1 firms respectively with

$$C(\overline{q}_1) \leq \sum_{i=2}^{n} C(\overline{q}_i). \tag{1}$$

For the water sector in a given geographical area, the above definition is true in the case of a single output. Indeed, sharing a pipeline network to supply water is very difficult without incurring excess costs due to inefficiencies associated with the sub-optimal utilization of productive capacity (double use for the same type of operation, over-capitalization, etc)[1]. Thus, in the case of drinking water management, for a given level of output, equation 1 is a strict inequality.

Over time, water management has evolved to the point where it requires the use of increasingly sophisticated knowledge and technologies related to a high level of very complex research and development[2]. This is because regulatory standards tend to evolve so that the community can benefit from the latest scientific know-how[3]. However, the latest scientific know-how is not easily transmissible without cost and remains the property of the industrial companies that carried out the research and hold the patents. As a consequence, the position of the industry that manages the sector is reinforced. Indeed, it becomes a de facto technological monopoly.

Thus, the current situation of the water sector can be characterized as a "double" monopoly. The first element relates to the specificity of the hydrous network which can be shared only at the price of resource wastage (work, capital). It corresponds to the traditional definition of a natural monopoly. The second element relates to the patent protected technological advances in water management that create temporary technological monopolies and reinforce the structure of the natural monopoly.

2.2 Analytical consequences

Knowing that water management is a natural monopoly does not automatically solve the management problems. Theory and practice have shown that monopolies are inefficient when compared with free competition. The problem is to manage the monopoly in the public interest while preserving the resource. In the case of water, the key questions are allocation and distribution with respect to regulatory constraints, pricing, current output and investment that will determine future output.

The answers to these questions are complicated by ethical considerations related the special status of water as indispensible to life itself. This limits the role of water as a purely economic good and implies a special set of conditions for the water monopoly that recognize that no one can be excluded because of price. This aspect is particularly sensitive in the developing countries where a significant fringe of the population does not have sufficient financial resources to pay the economic price for scarce water.

At this point the situation boils down to the determination of management criteria compatible with social criteria and the question of how the economic rents stemming from the monopoly situation can best be managed. The determination of management criteria fulfilling the objectives of economic efficiency and fair access to the resource constitutes the basis of relations between the regulator and the natural monopoly. The State (or the regulator) must guarantee the realization of both criteria: economic efficiency and social criteria for the community.

3. PRICING POLICIES: RANGE AND LIMITS OF THE STANDARD ANALYSIS

The foregoing discussion brings up the practical problem of how to determine a price for water that guarantees current and future quality, quantity and access for all with maximum efficiency. To this end we should make it clear what we mean by price.

According to the OCDE (1999, p.19), a tariff is a whole set of procedures and elements that determine the total water bill paid by the consumer; a tax is that part of the total bill measured in money per unit of time or in money alone; price is that part of the total bill measured in money per volume. Thus, in this context, the cost of water is a broader concept than the notion of price alone. However, to simplify the discussion, in the following paragraphs, we refer to the total bill paid by the consumer as the price.

3.1 Practices of water pricing

It is generally admitted that pricing in the water sector is largely the legacy of traditions and practices related to the country's history and sociology. Thus, the fixed price payment is still in practice, determined by the diameter of the drain (Austria), by the size of the meter (Australia), by surface living area (Norway), etc. However, the tendency is toward volumetric pricing combined with a fixed charge. The size of the fixed charge is the subject of discussion regarding equity and access. Experience has shown that higher fixed charges tend to reduce access and the consideration of social criteria. Current pricing practice is to take into account the marginal costs of supply related to volumetric consumption and fixed costs related to infrastructure investment. The effect of this practice on equity and access depends on the proportion of the fixed share compared to the price of the cubic meter. If the fixed share is too high, equity and access are reduced whereas if it is too low, the maintenance of existing or construction of new infrastructure is neglected.

3.2 Analytical basis of natural monopoly pricing (known as "Ramsey-Boîteux")

The already mentioned OCDE (1999) study shows how complex is the determination and application of new pricing structures. The presentation that follows constitutes a significant simplification compared with the effective practices associated with price formation. However, marginal cost pricing (as opposed to fixed cost pricing) is a popular method that appears to be generally well-adapted to answer the two objectives of economic effectiveness and social justice.

In the immediate aftermath of World War II, marginal cost pricing was discussed at length in the European administrations charged with determining the pricing policies of nationalized monopolies. In order to force the nationalized monopolies to respect their role of public utility services, i.e. maximizing the collective surplus, marginal cost pricing was enforced as a rule. However, this practice caused durable losses for some companies whose marginal costs were lower than their average costs.

Early work on natural monopoly pricing and management by Dupuit (1849), Hotelling (1938) and Vickry (1948) emphasized pricing at the marginal cost. This pricing scheme produces first rank optimum quantities only. Indeed, when considering the natural monopoly case, it appears that the marginal cost pricing leads to first order optimality from the quantity side. But, because scale returns are increasing, marginal costs are always lower than average costs. So the natural monopoly is incurring losses. The simplest way to remedy that is to get government subsidies. However this solution to make up the difference is also fraught with difficulties because the subsidies must be paid for with taxes, which create other distortions. There is also a problem of asymmetric information associated with the possible dichotomy between the effective costs and the costs declared by the company.

Faced with these difficulties, other pricing schemes have been proposed. Boiteux (1956) applied the work of Ramsey (1927) to the management of public companies and suggested that the price paid by the consumer should be such that the difference from the marginal cost should inversly proportional to the price elasticity of demand in order to cover the company's fixed costs. Known as Ramsey-Boiteux marginal cost pricing, it seeks the price compatible with public welfare and production constraints. More specifically, it aims at maximizing the collective surplus under the balanced budget constraint. This is a second best optimum where prices are higher than the effective individual willingness to pay. According to the Ramsey-Boiteux rule, the fundamental variable is the price-elasticity of demand for each good and service. The marginal cost is the pricing benchmark but when the budget constraint is taken into consideration, prices are settled at a different level.

One practical problem with Ramsey-Boiteux is that of assymetric information where the firm knows its true cost but the regulator does not[4]. Where water management is concerned there is another problem with Ramsey-Boiteux because it means that the willingness to pay is higher than the short-term marginal cost. However, in many poor and arid areas, the necessary infrastructures are so costly that for most of the population the willingness to pay is always less than the marginal cost. In these cases, another pricing method that reconciles community welfare and the budget constraint is required.

3.3 Two-part pricing

Two-part pricing, called equally non-linear pricing, is used by many public utilities. This pricing gives some freedom to redistribute the social surplus. The principle is relatively simple. The consumer pays a fixed charge (for instance a subscription) plus the bill of its effective consumption. If C is the fixed amount and c the unit price, the total bill S for consumption q is:

$$S = C + cq \tag{2}$$

This pricing scheme can insure a first order optimum if the fixed charge is set to offset the spread between marginal cost and average cost, which involves a deficit (D) when pricing to the marginal cost under increasing returns to scale. Roberts (1979) and Sharkey and Sibley (1993) show in a partial equilibrium framework that when a monopoly supplies to consumers a schedule of contracts, specifying the fee and the charge price, then the natural monopoly regulator can redistribute towards the weak demand consumers. In this case, the fixed charge is adjusted to the deficit (D) for a given number of N subscribers such that:

$$D = N\,C \tag{3}$$

This program can easily be generalized because natural monopolies tend to differentiate the fixed charge, called a subscription price, according the expected quantities to be consumed. Thus the problem of inequality can be overcome by charging a higher subscription price to the larger consumers. This is usually the case in practice where the subscription price increases with respect to the expected quantity of water to be consumed by the individual consumers. In this way, the pricing takes account of the social criteria. A kind of equity is then restored. The low quantity consumers pay a lower subscription price than the high quantity consumers.

3.4 Limits of the standard theories

Traditional pricing paradigms are based on a static approach and consider neither risk nor dynamic variables such as population growth, technological evolution, network improvement, etc. For example, the Ramsey-Boîteux approach reconciles welfare maximization and the budget constraint but only for one state of nature where population, technology network, etc. are all given and there is no risk such as quality deterioration or network failure[5]. Another problem is that they do not deal with the problem of natural "technological" monopolies discussed in paragraph 2 that make it possible for the private water companies holding the technology patents to extract economic rents at the expense of the community. Finally, they do not address the problem of what the private firm should pay the regulator for the right to exploit the monopoly. In the following section we incorporate the problem of "technological" monopolies and dynamize the pricing paradigm by proposing a model that includes time and risk. We then use the model to determine the optimal water price as well as the price that the firm should pay the regulator for the right to exploit the monopoly.

4. DYNAMIC PRICING WITH REAL OPTIONS

In a pair of papers, Clark and Mondello (2000a and b) introduce uncertainty with respect to water quality, quantity and network failure and look at monopoly pricing from the persepctive of privatization through delegation in the presence of "technological" monopolies. These monopolies refer to the patents that allow de facto cartels of large, specialized companies to control the technology of monitoring the safety of existing pipe networks and the skilled labor that is necessary to use the technology. Because of these monopolies, delegation of the management of the network to a private company that is contractually supposed to be for a limited period often becomes irreversible in practice.

In Clark and Mondello (2000a) we show that the risk of quality deterioration and network failure can be measured as the value of a hypothetical, infinitely lived insurance policy that pays off all losses due to quality deterioration and network failure. Exposure to loss is modeled as geometric Brownian motion, and loss causing events, including changes to higher regulatory quality standards, are modeled as Poisson processes. Thus, if the value of the hypothetical insurance policy is noted as Y, the annual premium l is equal to rY, where r is the risk-free interest rate, so that the present value of the premiums is just equal to the value of the policy. As we will see in the following paragraphs, this premium figures directly in the pricing formula when quantities to be supplied are uncertain.

4.1 The pure value of the monopoly

Clark and Mondello (2000b) consider delegation and the pricing problem when the quantities to be supplied to meet demand can vary a stochastically. This variation can be due to changes in population and social habits, the weather, public events, and the like. Gross income from the water project is equal to the price per unit, which is to be determined by the regulator, multiplied by the number of units supplied. Let x(t) represent gross income Since gross income cannot be negative, its evolution through time can be captured by geometric Brownian motion:

$$dx(t) = \alpha x(t)dt + \sigma x(t)dz(t) \tag{4}$$

where α is the expected growth rate of gross income, σ is the standard deviation of the growth rate and dz is a standard Wiener process with zero mean and variance equal to dt. If we make the standard Ramsey-Boîteux assumption that population, technology and network are given, α is equal to zero.

The required risk adjusted rate of return on $x(t)$ can be determined by applying the CAPM directly to $x(t)$ [6]. The required rate of return, noted as μ, will then be given by

$$\mu = r + \Lambda\sigma\rho_{x,m} \tag{5}$$

where r is the riskless rate of interest, Λ is the market price of risk, $\rho_{x,m}$ is the correlation coefficient of the percentage change in $x(t)$ with the market rate of return. The dividend or convenience yield, noted as δ, is equal to the difference between the required rate of return and the growth rate of x: $\mu - \alpha = \delta$ Under the the standard Ramsey-Boîteux assumption of population, technology and network as given, α equals zero and, thus, $\delta = \mu$.

Let c represent operating costs that include the insurance premium covering losses due to quality deterioration and network failure discussed above. To simplify the analysis, we assume that within the output capacity of the investment operating costs are constant[7]. Thus, net income is equal to $x(t) - c$. We can use this information to calculate the pure value of the monopoly. The pure value of the monopoly refers to the value of the investment based purely on the cash flows it can generate at the optimal price. As discussed above, the essential nature of water in all facets of human activity makes water management and supply a requirement for any organised society. It is an indispensable activity that must exist as long as the society itself exists. In this sense, water management and supply do not depend explicitly on time and, for all practical purposes, can be considered as an infinitely lived investment. With this in mind, the pure value of the monopoly, noted as $V(x(t))$, can be found by setting up a hedge portfolio with a long position of one unit of the investment and a short position in $V'(x(t))$ units of $x(t)$. Using standard methods in stochastic calculus gives the following differential equation[8]:

$$\frac{\sigma^2}{2}V''(x(t))x(t)^2 + (r-\delta)V'(x(t))x(t) - rV(x(t)) + x(t) - c = 0 \tag{6}$$

where the primes represent first and second derivatives.

The solution to (6) is:

$$V = \frac{x(t)}{\delta} - \frac{c}{r} + A_1 x(t)^{\gamma_1} + A_2 x(t)^{\gamma_2} \tag{7}$$

where $\gamma_1 > 1$ (because $\delta > 0$) and $\gamma_2 < 0$ (because $r > 0$) are the roots to the quadratic equation in γ:

$$\gamma_{1,2} = \frac{-(r-\delta-\frac{\sigma^2}{2}) \pm \sqrt{(r-\delta-\frac{\sigma^2}{2})^2 + 2\sigma^2 r}}{\sigma^2} \tag{7a}$$

The constants A_1 and A_2 depend on the boundary conditions. The first boundary condition in straightforward. When income is equal to zero, the investment has no value:

$$V(0) = 0 \tag{8}$$

This condition implies $A_2 = 0$. If we rule out speculative bubbles, the second boundary condition is:

$$V'(\infty) < \infty \tag{9}$$

which implies that $A_1 = 0$. Thus, the solution to (7) is:

$$V = \frac{x(t)}{\delta} - \frac{c}{r} \tag{10}$$

Equation 10 says that the pure value of the monopoly is equal to the present value of the net cash flows where income (x) is discounted at the risk adjusted rate because it is uncertain and cost (c), because it is constant, is discounted at the riskless rate.

4.2 The regulator's option to revoke delegation

When water management is delegated to a private firm, the regulator's position changes from holding the monopoly himself to holding the right to revoke delegation and take the monopoly back from the delegated firm. On the other hand, the delegated firm holds the monopoly and is now the monopolist[9], but it has the right (de facto or de jure) to renounce its delegation contract (abandon the project) if it so desires. Both of these rights are options and can be evaluated as such. The regulator's right to reverse delegation is contractual and can either be American style where the decision can be made at any time or European style where the decision can be made only on certain dates. The firm's right to abandon can be contractual but it is ultimately de facto in so far as in practice the firm can pay the indemnities and abandon the contract or just go bankrupt. It can also be either American or European style. For simplicity of exposition, we model both options as American style. As we show below, the optimal water price and the price the delegated firm must pay for the rights to the monopoly depend on the values of these two options. We start with the regulator's option to revoke delegation.

Let $F = F(x(t))$ represent the value of the regulator's option that gives him the right to revoke delegation at the exercise price I. This exercise price includes the technology costs, recruiting costs, investment costs and indemnities that must be paid if the regulator wants to renew direct management. The regulator will only want to renew direct management if it is in his interest to do so, that is, the level of income must be high enough to offset the costs of revoking the delegation. The level of income that will trigger revocation of the delegation, noted as x^*, is found by solving the differential equation

$$\frac{\sigma^2}{2} F''(x(t))x(t)^2 + (r - \delta)F'(x(t))x(t) - rF(x(t)) = 0 \tag{11}$$

under the boundary conditions

$$F(0) = 0 \tag{12}$$

and

$$F(x^*) = V(x^*) - I \tag{13}$$

and the smooth pasting condition that makes it possible to find x^* jointly with $F(x(t))$ [10]:

$$F'(x^*) = V'(x^*) \tag{14}$$

Equation (12) means that when the investment has no value, the option to revoke has no value either. Equation (13) means that at the optimal level of income x^*, the value of the option that the regulator gives up is just equal to the value he receives, that is, the pure value of the monopoly less I, the price he must pay to revoke. Equation (14) is a technical condition, called the smooth pasting condition that rules out arbitrage around the exercise point.

This gives:

$$F = B_1 x(t)^{\gamma_1} \tag{15}$$

where $\gamma_{1,2}$, the roots to the quadratic equation in γ, are given in (7a),

$$B_1 = \frac{(\gamma_1 - 1)^{\gamma_1 - 1}\left[\frac{c}{r} + I\right]^{1-\gamma_1}}{(\delta\gamma_1)^{\gamma_1}} \tag{16}$$

and

$$x^* = \frac{\gamma_1}{\gamma_1 - 1}\delta\left[\frac{c}{r} + I\right] \tag{17}$$

Equation (17) gives the maximum income level for the monopolist firm that is acceptable to the regulator. Beyond this point, it is in the interest of the regulator to revoke delegation and resume direct management. This solution includes the risk of variations in cash flows (σ) as well as the risk of damage liability, which is subsumed as the insurance premium in c, the operating costs. The two types of risks have different effects on x*. Changes in σ reduce x* ($\partial x^*/\partial\sigma < 0$), the maximum level of income that the regulator can allow the monopolist firm, whereas changes in the insurance premium increase it ($\partial x^*/\partial c > 0$). Thus, more uncertainty regarding cash flows reduces the monopolist firm's scope for realizing abnormal profits. However, increases in damage liability do not necessarily raise the scope for economic rents. Although increases in damage liability raise the regulator's acceptable maximum price, they also raise the monopolist's costs. Thus, it appears that moral hazard is reduced since it is in the interest of both parties to work to reduce the risk of losses due to water quality and network failure[11]. The regulator wants to reduce accidents and protect the interests of the community and the monopolist wants to reduce his costs. This he can do by upgrading his technology, which probably goes a long way to explaining the rapid pace of technological improvement observed in France since delegation became popular. Of course, investments in technology could also raise the monopolist's operating costs in the form of depreciation, so there is a trade-off here, beyond the scope of this paper, between risk reduction and depreciation costs.

4.3 The monopolist firm's option to abandon delegation

As we mentioned above, the monopolist can abandon delegation if it feels that the project is no longer worth operating. The regulator does not have this option, since by law it is obliged to assure the supply of water to the area it administers. Thus, although from the regulator's standpoint the monopoly is, for all practical purposes, an infinitely lived investment, as a simple agent, the monopolist firm is in the position of being able to terminate the contract and turn water management back over to the regulator if delegation is no longer in its interest. This will be the case if income falls too low. To determine the value of this abandonment option and the level of income where abandonment is advantageous, we proceed as before.

Let $Z(x(t))$ represent the value of the investment that includes the abandonment option. Next, build a hedge portfolio consisting of one unit of the investment $Z(x(t))$ and a short position of $Z'(x(t))$ units of income. Going through the same steps as before gives the following differential equation:

$$\frac{\sigma^2}{2}Z''(x(t))x(t)^2 + (r-\delta)Z'(x(t))x(t) - rZ(x(t)) + x(t) - c = 0 \quad (18)$$

whose solution is:

$$Z = \frac{x(t)}{\delta} - \frac{c}{r} + D_1 x(t)^{\gamma_1} + D_2 x(t)^{\gamma_2} \quad (19)$$

The only difference between (19) and (7) are the constants. In the absence of speculative bubbles, as $x \to \infty$, the value of the abandonment option goes to 0 and, therefore, $D_1 = 0$. For the value matching condition, define S, the net salvage value from the delegation contract, as equal to gross salvage value less the abandonment costs. Gross salvage value is basically contractual and comprised of the undepreciated value of the firm's investment in infrastructure and technology. Abandonment costs are also basically contractual and include cash penalties as well as certain costs associated with assuring the transition from one management to another. Since the abandonment option is a put, net salvage value represents the exercise price[12].

Thus, there will be a level of x, noted as x^{**}, that is so low that the monopolist would be better off abandoning the monopoly to the regulator. At this point, the value matching condition is $Z(x^{**}) = S$ and the smooth pasting condition is $Z'(x^{**}) = 0$.

Solving for x^{**} and D_2 gives:

$$x^{**} = \left[S + \frac{c}{r}\right]\frac{\delta\gamma_2}{\gamma_2 - 1} \tag{20}$$

where $x^{**} > 0$ because $\delta > 0$ and $\gamma_2 < 0$

$$D_2 = -\frac{x^{**1-\gamma_2}}{\delta\gamma_2} \tag{21}$$

This gives the position of the monopolist as

$$Z(x(t)) = \frac{x(t)}{\delta} - \frac{c}{r} + D_2 x(t)^{\gamma_2} \tag{22}$$

From (22) we can see that the monopolist's position is equal to the pure value of the monopoly $V(x(t)) = \frac{x(t)}{\delta} - \frac{c}{r}$ plus the value of the option to abandon the monopoly $D_2 x^{\gamma_2}$. Since the abandonment option is a put, its value increases as the exercise price increases. This can easily be verified by taking the first partial derivative of the option with respect to S.

4.4 Price determination

In this section, we suggest how the foregoing discussion can be used to determine a fair deal for both the regulator and the monopolist firm. The fair deal should eliminate economic rents while rewarding the risk taken on by the firm. It should also guarantee that the regulator receives the full value of the monopoly that it is ceding to the firm. In this sense the fair deal is an optimum. To arrive at this solution, we look at the net position of the monopolist firm.

The net position of the monopolist firm is equal to the pure value of the monopoly plus the abandonment option (Z) less the value of the call option (F) that was effectively issued to the regulator when the firm accepted the delegation contract. This gives

$$Z - F = \frac{x(t)}{\delta} - \frac{c}{r} + D_2 x(t)^{\gamma_2} - B_1 x(t)^{\gamma_1} \tag{23}$$

From equation 10 we know that the pure value of the monopoly is equal to the present value of the net cash flows. From (23) we can see that this will be the case when $D_2 x^{\gamma_2} = B_1 x^{\gamma_1}$ where S and I represent economic values as opposed to values based on the technological monopoly of the delegated firm. Thus, a fair deal for both firm and regulator can be achieved by determining the economic values of S and I, observing the quantity of water currently being consumed, and then setting the price per unit that, when multiplied by the quantity currently being consumed, gives the gross income x that equalises the two option values[13]. With x thus determined, the pure value of the monopoly that the regulator is ceding to the firm is also determined. To complete the fair deal, the delegated firm would then pay the pure value of the investment for the right to exploit the monopoly. The total outcome for both parties would then be zero such that neither party profits at the expense of the other. This can be seen by an example.

Consider the following information summarized in table 1 where the value of S includes \$15 in net investment, \$2 in training and hiring and \$3 in cash penalties while I has the same figures for net investment and training and hiring and \$8 for the cost of procuring the new technology. Suppose that all these costs except penalties represent the fair market prices[14].

Table 1: Parameters information

α	μ	r	σ	S	I	c
0	8%	5%	10%	\$15 – \$2 –\$ 3 = \$10	\$15 +\$ 2 + \$8 = \$25	\$1

Using this information in equations 7a, 16, 17, 20 and 21, we find that $\gamma_1 = 8.2167$, $\gamma_2 = -1.217$, $B_1 = 0.0000572$, $D_2 = 18.9266$, $x^* = 4.0988$ and $x^{**} = 1.31745$. The value of x that gives $D_2 x^{\gamma_2} = B_1 x^{\gamma_1}$ is \$3.8436. Thus, the optimal gross income level should be set at \$3.8436. The regulatory price can then be deduced from the current quantity of water being consumed. Suppose that 4 units are currently being consumed. The unit price that gives the optimal income of \$3.8436 is \$0.9609. This price includes the uncertainty with respect to quantity reflected in σ as well as the risk of quality deterioration and network failure reflected in the insurance premium. In these conditions, the amount that the delegated firm would have to pay for the right to exploit the monopoly would be equal to the pure value of the monopoly estimated at $x = \$3.8436$, which in the present case works out to $\$3.8436/0.08 - \$1/0.05 = \$28.045$.

5. CONCLUSION

In this paper, we have outlined some of the difficulties associated with the supply and pricing of drinking water. We also reviewed the standard pricing approaches, Ramsey-Boiteux marginal cost pricing and non-linear pricing, and outlined their limitations, notably that they are static and relevant to only one state of nature. They also fail to deal with the important question of price that the private firm should pay for the right to exploit the monopoly. We then developed a model based on standard techniques in real option theory that overcomes these difficulties and that can be used as a pricing program that gives a fair deal to both the regulator and the delegated firm. Our pricing program incorporates the fact that water supply conforms to a natural monopoly complicated by ethical and moral considerations, which are due to the special nature of water as a basic requirement for human life. It goes beyond marginal cost and non-linear pricing by introducing time and risk and provides a solution to the price that the delegated firm should pay for the right to exploit the monopoly. It also includes the relatively recent problem of technological monopolies that make it possible for the private water firms to extract supplementary economic rents. The fair deal in our pricing program eliminates economic rents while rewarding the risk taken on by the firm. It also guarantees that the regulator receives the full value of the monopoly that it is ceding to the firm. In this sense the fair deal that we propose is an optimum.

Notes

1. John Stuart Mill inuitively shared this analysis when he wrote: "It is obvious, for example, that one could save much work if London were supplied by only one gas or water company rather than by the existing plurality".

2. Brubaker, (2002), see also the report of Boyer et al. (1996), Garcia and Thomas (2003), Clark and Mondello (2002).

3. For instance, for United States, the Safe Drinking Water Act (SDWA), which celebrates its 30th anniversary in 2004, is the main federal law that ensures the quality of Americans' drinking water. Under SDWA, EPA sets standards for drinking water quality and oversees the states, localities, and water suppliers who implement those standards.

We can refer to the EU Council Directive 98/83/EC on the quality of water intended for human consumption that advocates improving the quality rules and adopted the following main changes in parametric values:

- Lead: reduced from 50 µg/l to 10 µg/l, 15 years transition period to allow for replacing lead distribution pipes,
- Pesticides:values for individual substances and for total pesticides retained (0.1µg/l / 0.5µg/l), plus additional, more stringent ones introduced for certain pesticides (0.03µg/l),
- Copper: value reduced from 3 to 2 mg/l ,
- Standards introduced for new parameters like trihalomethanes, trichloroethene and tetracholoroethene, bromate, acrylamide etc.

4. See, for example, Baron and Myerson (1982) and Laffont and Tirole (1986). They analyse the problem in the principal-agent paradigm where the regulator is the principal and the company is the agent.

5. See Boyer and Robert (1997) for a review of the informational difficulties of the standard approaches.

6. In many countries, such as France, $x(t)$ is directly observable in so far as water companies are required by law to furnish the authorities with regular, detailed information on quantities and prices. If $x(t)$ were not directly observable, a spanning asset could be substituted. An alternative method in the absence of a reliable spanning asset is to assume risk neutrality.

7. This assumption implies that either there is a single given technology or that technology changes affect water quality but not cost. Technology and changes in technology, while important in a more general context, are only peripheral to the problem at hand. Furthermore, the great majority of operating costs in water management accrue to depreciation, which is fixed or has a fixed schedule, and skilled labor. Because of social legislation that eliminates temporary layoffs and the specialized skills required for water management that make temps a limited commodity, labor costs are also fixed for all practical purposes. Thus, variable operating costs are a small percentage of total operating costs. Consequently, the assumption of constant operating costs is not too unrealistic. Since this assumption simplifies the mathematics and makes the model more intuitively appealing, we have much to gain and little to lose by it.

8. For a presentation of this technique commonly used in Financial modelling, see any textbook using stochastic calculus such as Dixit and Píndyck (1994).

9. In fact, it is a double monopolist. It has the monopoly rights to the network as well as the monopoly rights to the technology it employs.

10. For details of the techniques involved see Dixit and Pindyck (1994).

11. This involves reducing the value of the hypothetical insurance policy by reducing the exposure to loss, the probality of loss causing events or some combination ogf the two.

12. The firm gives up the project and receives the salvage value.

13. See Clark and Mondello (2000b, pp. 343-348) for a discussion of the practical difficulties and how they can be overcome.

14. See Clark and Mondello (2000b pp. 346-348) for a discussion of how the market prices could be estimated.

References

Averch H. and L. Johnson, 1962. 'Behavior of the firm under regulatory constraint', *American Economic Review*, 52, 1052-1069.

Baron, D. P. and R. B. Myerson, 1982. 'Regulating a monopolist with unknown costs', *Econometrica*, 50(4), 911-930.

Baumol, W., J. Panzar and R. Willig, 1982. *Contestable Markets and the Theory of Industry Structure* (revised edition), Harcourt Brace Jovanovich, 1988.

Becker, G.S., 1983. 'A theory of competition among pressure groups for political influence', *Quarterly Journal of Economics*, 98, 371-400.

Boiteux, M., 1956. 'Sur la gestion des monopoles publics astreints à l'equilibre budgétaire', *Econometrica,* 24, 22-40.

Boyer, M. and J. Robert, 1997. 'Competition and Access in Electricity Markets', ECPR, Global Price Cap, and Auctions, CIRANO, Scientific Series, Montreal.

Boyer M., P. Lasserre, M. Patry, M. Poitevin and B. Sinclair-Desgané, 1996. 'L'impartition municipale : l'argumentaire', Rapport CIRANO, Montréal, june, 70 pages.

Brubaker, E., 2002. 'Liquid assets: privatizing and regulating Canada's water utilities', University of Toronto Centre for Public Management.

Clark, E. and G. Mondello, 2000a. 'Resource management and the Mayor's guarantee in French water allocation', *Environmental and Resource Economics*, 15(2), 103-113.

Clark, E. and G. Mondello, 2000b. 'Water management in France: delegation and irreversibility', *Journal of Applied Economics*, 3(2), 325-352.

Clark, E. and G. Mondello, 2002. 'Regulating natural monopolies: the case of drinking water in France', *Water Resources Update*, issue no. 121, (January 2002): 72-78.

Clark, E. and G. Mondello, 2003. 'Water management in France: delegation and market based auto-regulation', *International Journal of Public Administration,* Special issue, 26 (3).

Dixit, A. and R. Pindyck, 1994. *Investment Under Uncertainty*, Princeton, NJ, Princeton University Press.

Dupuit, J., 1849. 'De l'influence des péages sur l'utilité des voies de communication', *Annales des Ponts et Chaussées*, 2e série, 170-248.

Garcia, S. and A. Thomas, 2003. 'Regulation of public utilities under asymmetric information: The case of municipal water supply', France, *Environmental and Resource Economics*, 26, 145–162.

Hotelling, H., 1938. 'The general welfare in relation to problems of taxation and of railway and utility rates', *Econometrica*, 6(3), 242-269.

Laffont, J.-J. and J.Tirole, 1986. 'Using cost observation to regulate firms', *Journal of Political Economy*, 94(3), 614-641.

Lyman, R. A., 1992. 'Peak and off-peak residential water demand', *Water Resources Research*, 28 (9), 2159-2167.

Mill, J.S., 1851. *The Regulation of the London Water Supply* , reprinted in Essay's on Economics and Society, Collected Writing.

OCDE, 1999. 'Tarification de l'eau à usage ménager dans les pays de l'OCDE', ENV/EPOC/GEEI(98)12/FINAL, 77843.

Peltzman, S., 1976. 'Toward a more general theory of regulation', *Journal of Law and Economics*, 19, 211-240.

Pezzey, J. and G. Mill, 1998. A Review of Tariffs for Public Water Supply: A Report to the Enviroment Agency. National Water Demand Management Centre, Worthing.

Ramsey, F. P., 1927. 'A contribution to the theory of taxation', *Economic Journal*, 37, 47-61.

Roberts, K., 1979. 'Welfare considerations of non-linear-pricing', *Economic Journal*, 89, 66-83.

Stigler, G., 1971. 'The theory of economic regulation', *Bell Journal of Economics and Management Science,* 2, Spring, 3-21.

Sharkey W. and D. Sibley, 1993. 'Optimal non-linear pricing with regulatory preference over customer type', *Journal of Public Economics*, 50, 197-229.

Vickrey, W., 1948. 'Some objections to marginal-cost pricing', *Journal of Political Economy*, 56, 218-238.

Chapter 5

PRICE RISK AND THE DIFFUSION OF CONSERVATION TECHNOLOGY*

Georgina Moreno
Scripps College, U.S.A.

David Sunding
University of California, Berkeley, U.S.A.

1. INTRODUCTION

In recent years, technology diffusion has become a main theme of environmental economics and the economics of public utility regulation. This emphasis stems from the demonstrated ability of new technologies to ameliorate the conflict between environmental quality and economic activity. Indeed, some economists have argued that the successful diffusion of new technologies is among the most important factors leading to the success of environmental policies and improvement in environmental quality (Kneese, 1978).

Conservation technologies are prime examples of technology with the potential to benefit the environment. These technologies typically entail an up-front investment (which is often substantial), but produce a given level of output with a lower level of factor use. Examples abound, and include energy-conserving heating and cooling systems, fuel-efficient cars, and water-saving appliances and irrigation systems.[1]

*This research was funded by the U.S. Environmental Protection Agency and the California Resources Agency under the CALFED Program, and by a Challenge Grant from the U.S. Department of the Interior. The opinions expressed here are not necessarily those of the funding agencies. The authors thank Michael Hanemann, David Zilberman, Kenneth Train, Daniel McFadden, Elizabeth Sadoulet and Jeff Perloff for helpful comments, and also acknowledge participants in seminars at UC Berkeley, UC Santa Barbara and the USEPA.

Economic models of technology diffusion have established the principle that relative prices have a large effect on the adoption decision (Sunding and Zilberman, 2001). Existing empirical studies of conservation technology show that these technologies offer no exception to this general principle. In particular, a number of economists have established that the diffusion of conservation technology is stimulated by an increase in factor prices. A recent study by Pizer et al. (2002) examined the highly polluting oil refining, plastics, pulp and paper, and steel industries and found that increases in energy prices increased the likelihood of adopting conservation technologies. Rose and Joskow (1990) demonstrated that fuel price increases had a positive and significant effect on the diffusion of a fuel-saving technology in the electricity industry. Similarly, Boyd and Karlson (1993) showed that the diffusion of an energy-saving technology used in the U.S. steel industry was positively related to the price of fuel.

While the effect of input price on adoption is fairly clear, the influence of changes in factor price risk on long-run efficiency is not as well understood. This omission in the literature is significant since price volatility is quite evident in resource markets, and since some economists have speculated that price risk may play an important role in determining input-use efficiency. There have been some forays into this area, mostly focusing on the fact that the influence of price risk on option value and the timing of investment in conservation technology (see the survey article by Jaffe et al., 2001). There is undoubtedly merit to this argument, and empirical studies have confirmed, consistent with the option value hypothesis, that adoption of conservation technology is more likely to occur in periods when input price is high (see, for example, Carey and Zilberman, 2002). In this chapter we leave timing considerations aside and focus on the question of how factor price risk influences the expected returns from investment in conservation technology. Somewhat surprisingly, we find that an increase in risk has an ambiguous effect on the incentives to install conservation technology (and hence on long-run factor-use efficiency) since we identify realistic cases in which an increase in factor price risk makes the conservation technology less attractive relative to conventional ones.

We pursue the question of how factor price risk influences the diffusion of conservation technology by developing a conceptual model of long-run input-use efficiency. This framework is consistent with the seminal approach of Hausman (1979) that views technology adoption as embedded in a two-stage process of input demand. In the first stage, an input-use technology is selected, thereby fixing the input-output ratio. In the second stage, agents choose the level of output, which implies a level of factor utilization.

Our approach to the relationship between price risk and conservation technology adoption is also related to Abel (1983) on optimal factor intensity. Abel considers a mean-variance model of the choice of optimal factor intensity and shows that, in general, a marginal increase in the variance of the factor price has an ambiguous effect on factor intensity. While our chapter and Abel's are related in that they both use a two-stage framework, our conceptual model relies on a more general stochastic dominance approach instead of a change in variance in a two-moment model as considered by Abel. In particular, our conceptual analysis considers the case of a mean-preserving increase in factor price risk, which is perhaps the purest expression of an increase in risk. Economists have been aware for some time of the distinction between an increase in risk and an increase in variance. Although this notion is usually attributed to Rothschild and Stiglitz (1970), the same point was made by Borch (1969), and was appreciated decades earlier by some researchers outside economics (for example, Hardy et al., 1934).

A main conclusion of our conceptual analysis is that a mean-preserving increase in price risk can increase or decrease the incentive to adopt conservation technology, and thus has an ambiguous effect on the optimal level of factor-use efficiency. The direction of the change in efficiency relates to the short-run elasticity of input demand with respect to the input price, that is, the elasticity of factor demand conditional on the status quo technology. If factor demand under the existing technology is elastic with respect to the input price, then an increase in risk increases the return from investment in more conservation technology and thus increases the optimal level of factor-use efficiency. If utilization is inelastic, then the opposite result holds. The intuition for the result hinges on the relation between the magnitude of the elasticity of factor demand and the concavity or convexity of the expected profit function with respect to the factor price, as will be explained in the next section.

The conceptual model is then used to formulate an econometric test of the influence of price risk on factor-use efficiency. We utilize a unique data set from the water industry comprised of observations on water use efficiency at the micro level for two groups of farms served by the same water utility. By design, these groups face prices with identical means but different levels of price risk. Assignment to one of the groups is determined by historical water rights that are appurtenant to the land and not the owner. We estimate an ordered probit model of technology choice whose form is determined by the conceptual model. In particular, we estimate a model in which factor price risk is interacted with a measure of the elasticity of utilization. Results are consistent with the main hypotheses, and highlight the importance of considering the impact of changes in price risk at an appropriate level of disaggregation.

The chapter is structured as follows. In Section 2, we present the basic conceptual model. Section 3 lays out the main conceptual results concerning the effect of price risk on conservation technology adoption. Section 4 presents the empirical model, data and estimation results. Concluding comments relating to policy implications, aggregate input-use efficiency and induced innovation are given in Section 5.

2. CONCEPTUAL MODEL

Consider the input use problem of an individual agent such as a household or firm. Following Hausman (1979), we suppose that water use per period is the result of a two-stage decision process. In the first stage, the agent chooses the level of water use efficiency (measured by the input-output coefficient,a), by selecting among a continuum of possible technologies. Each possible technology is characterized by the pair (a, z), where z is the annualized fixed cost of the technology. More efficient technologies require a larger outlay, and thus $z = z(a)$ with $z' < 0$, where z' is the derivative of z with respect to a.

In each period, the agent chooses an output level, x. The consumer's periodic water use is then $\overline{a}x$, where the bar denotes that the efficiency of water use is fixed by the prior choice of technology. The agent faces a stochastic water price, $p \in [\underline{p}, \overline{p}]$. The parameter p has a known distribution $F(p, \theta)$, where θ indexes risk and the corresponding density function is $f(p, \theta) = dF(p, \theta)$.

This model of input use and technology choice can be applied to a number of different energy and natural resource goods. For example, in the case of air conditioners, the activity level x is degree-hours cooled, p is the price of electricity, and a is the electricity requirement per unit of cooling. In the case of automobiles, x is the number of miles driven, p is the price of gasoline, and a is the inverse of miles per gallon. In both cases, consumers decide input-use efficiency through their choice of technology based on their expectations about factor prices, and choose the level of factor use in each period based on the previous choice of input-use efficiency and the current factor price.

Now consider the conditions for short- and long-run optimization. In the short-run, the input-output ratio is fixed and the agent chooses the activity level to maximize welfare conditional on known prices. The agent's short-run optimization problem is given by

$$\max_{x} \quad U = B(x) - pax - z(a) \tag{1}$$

subject to

$$a = \overline{a}$$

where $B(x)$ are benefits derived from output level x, $B' > 0$ and $B'' < 0$. The first order condition for this problem is

$$B' - p\overline{a} = 0 \quad \forall p, \tag{2}$$

which implicitly defines the optimal level of output as a function of the input price and the prior choice of technology, or $x = x(p, \overline{a})$. From condition (2), we obtain the following results:

$$x_a = \frac{p}{B''} < 0 \text{ and } x_p = \frac{a}{B''} < 0 \tag{3}$$

where x_a denotes the partial derivative of x with respect to a, and x_p denotes the partial derivative of x with respect to p. Denote the elasticity of water demand as $\epsilon = x_p p/x$. For simplicity, we consider the case of constant elasticity with respect to the factor price (but not with respect to a, however); later, we relax this assumption and show that the main result is only slightly modified.

Now we turn to the long-run investment decision in which the agent chooses the water-use technology to maximize expected utility. The long-run problem can be expressed as

$$\max_a \quad EU = \int_{\underline{p}}^{\overline{p}} [B(x) - pax - z(a)] \, f(p, \theta) dp \tag{4}$$

subject to

$$x = x(p, a).$$

The first-order condition for the technology choice problem is

$$\int_{\underline{p}}^{\overline{p}} \left[B'(x) - pax_a\right] f(p, \theta) dp - z'(a) = 0. \tag{5}$$

The first two terms in the integral cancel out by the short-run first-order condition (i.e., by the Envelope Theorem since $B'(x) - pa = 0 \; \forall p$). Thus, (5) simplifies to

$$-\int_{\underline{p}}^{\overline{p}} px f(p, \theta) dp - z'(a) = 0. \tag{6}$$

This optimality condition sets the expected marginal value of conservation, that is, the expected water cost savings from increasing efficiency, equal to the marginal cost of a more efficient technology. We use this condition to evaluate the impact of an increase in water price risk on the choice of water-use technology.

3. FACTOR PRICE RISK AND TECHNOLOGY CHOICE

We consider an increase in risk of the type described by Rothschild and Stiglitz (1970), namely a mean-preserving increase in risk.

DEFINITION: An increase in factor price risk occurs when

i) $\int_{\underline{p}}^{p} F_\theta(p, \theta) dp \geq 0 \; \forall p$, and

ii) $\int_{\underline{p}}^{\overline{p}} F_\theta(p,\theta)dp = 0$

where F_θ is the partial derivative of $F(p,\theta)$ with respect to θ.

It is now possible to show the following:

PROPOSITION: If an increase in θ increases risk as defined, then

$$sign\left[\frac{da}{d\theta}\right] = sign\left[|\epsilon| - 1\right]$$

Proof. Totally differentiate (6) with respect to a and θ and rearrange as follows:

$$\frac{da}{d\theta} = \frac{\int_{\underline{p}}^{\overline{p}} px f_\theta(p,\theta)dp}{LRSOC}. \tag{7}$$

LRSOC is the second-order condition of the technology choice problem, and is negative. Thus,

$$sign\left[\frac{da}{d\theta}\right] = sign\left[-\int_{\underline{p}}^{\overline{p}} px f_\theta(p,\theta)dp\right] \tag{8}$$

Integrating the right-hand side of (8), we see that

$$-\int_{\underline{p}}^{\overline{p}} px f_\theta(p,\theta)dp = -\overline{p}x(\overline{p},a)F_\theta(\overline{p}) + \underline{p}x(\underline{p},a)F_\theta(\underline{p}) + \int_{\underline{p}}^{\overline{p}} (x + px_p)\, F_\theta(p,\theta)dp. \tag{9}$$

The first two terms on the right-hand side of (9) equal zero since the supports of the factor price density are invariant. This leaves the expression

$$-\int_{\underline{p}}^{\overline{p}} px f_\theta(p,\theta)dp = \int_{\underline{p}}^{\overline{p}} (x + px_p)\, F_\theta(p,\theta)dp. \tag{10}$$

Integrate the RHS of equation (10) again to obtain

$$-\int_{\underline{p}}^{\overline{p}} px f_\theta(p,\theta)dp = [x(\overline{p},a) + \overline{p}x_p(\overline{p},a)] \int_{\underline{p}}^{\overline{p}} F_\theta(p,\theta)dp$$

$$- [x(\underline{p},a) + \underline{p}x_p(\underline{p},a)] \int_{\underline{p}}^{\overline{p}} F_\theta(p,\theta)dp$$

$$- \int_{\underline{p}}^{\overline{p}} \left[\int_{\underline{p}}^{p} F_\theta(p,\theta)dp \right] (2x_p + px_{pp})\, dp. \quad (11)$$

The first two terms on the right-hand side of (11) vanish: the first by the definition of the mean-preserving increase in risk and the second by the invariance of $\underline{p}$. Thus, we are left with the following expression:

$$-\int_{\underline{p}}^{\overline{p}} px f_\theta(p,\theta)dp = -\int_{\underline{p}}^{\overline{p}} \Psi(p) \left[\frac{x}{p}\epsilon\,(1+\epsilon) \right] dp, \quad (12)$$

where $\Psi(p) = \int_{\underline{p}}^{\overline{p}} F_\theta(p,\theta)dp \geq 0\ \forall p$ and we use the fact that

$$2x_p + px_{pp} = \frac{x}{p}\epsilon\,(1+\epsilon)\,, \quad (13)$$

which we obtain by using the assumption of constant elasticity with respect to price, thus ϵ is invariant with respect to p and $x_p = \epsilon(x/p)$. Therefore,

$$\begin{aligned} 2x_p + px_{pp} &= 2\frac{x}{p}\epsilon + p\frac{d}{dp}\left(\frac{x}{p}\epsilon\right) \\ &= 2\frac{x}{p}\epsilon + p\epsilon\frac{d}{dp}\left(\frac{x}{p}\right) \\ &= \frac{x}{p}\epsilon\,(1+\epsilon)\,. \end{aligned}$$

It follows that

$$sign\left[\frac{da}{d\theta}\right] = sign\left[|\epsilon| - 1\right]$$

as claimed.

The proposition establishes that the impact on technology choice of an increase in risk depends on the elasticity of input use with respect to the factor price, or more precisely, the elasticity of factor demand conditional on $a = a^*$.

To gain some intuition for this result, first note that $MP = px$ is the marginal productivity of an increase in water-use efficiency, or a decrease in the input-output ratio. Accordingly, $MP_{pp} = \frac{x}{p}\epsilon(1 + \epsilon)$ is the second derivative of marginal productivity with respect to the price of water. If this derivative is positive (i.e., if $|\epsilon| > 1$), then the marginal productivity of increasing the input-output ratio is convex in the input price, and a mean-preserving change in the factor price density increases the *expected* productivity of an improvement in input-use efficiency.[2] Conversely, if $|\epsilon| < 1$, then MP is concave and the opposite holds.

Because the marginal productivity of investment in conservation technology is quadratic in the input price elasticity, it is impossible to say exactly how an increase in elasticity will affect demand for the technology beyond the claim in the proposition. Nonetheless, even within the region where input demand is inelastic, there should be some relationship between demand elasticity and the effect of an increase in risk. This proposition is tested later in this chapter.

It is worth noting that the proposition is unaffected by the assumption that the agent never "shuts down"(i.e., does not use any factor at all if the price is too high). In fact, there may exist a factor price $\hat{p}$ such that $x(p,a) = 0\ \forall p \geq \hat{p}$. In this case, equation (9) becomes

$$-\int_{\underline{p}}^{\hat{p}} px f_\theta(p,\theta)dp = -\hat{p}x(\hat{p},a)F_\theta(\hat{p}) + -\underline{p}x(\underline{p},a)F_\theta(\underline{p}) + \int_{\underline{p}}^{\hat{p}} \left[x + px_p\right] F_\theta(p,\theta)dp. \quad (14)$$

The first term on the right-hand equals zero by the definition of $\hat{p}$. The second term equals zero since the supports of the density are invariant with respect to a change in risk. Thus, we are left with (10) as before (with the exception that the upper limit of integration is now $\hat{p}$). Then, (11) becomes

$$-\int_{\underline{p}}^{\hat{p}} px f_\theta(p,\theta)dp = [x(\hat{p},a) + \hat{p}x_p(\hat{p},a)] \int_{\underline{p}}^{\hat{p}} F_\theta(p,\theta)dp$$

$$+ \left[x(\underline{p},a) + \underline{p}x_p(\underline{p},a)\right] \int_{\underline{p}}^{\underline{p}} F_\theta(p,\theta)dp$$

$$- \int_{\underline{p}}^{\hat{p}} \left[- \int_{\underline{p}}^{p} F_\theta(p,\theta)dp \right] (2x_p + px_{pp}) \quad (15)$$

The first term on the right-hand side vanishes by the definition of $\hat{p}$ (i.e., since $x(\hat{p},a) + \hat{p}x_p(\hat{p},a) = x(\hat{p},a)\hat{p}(1+\epsilon) = 0$ at $\hat{p}$), and the second term vanishes because $F_\theta(\underline{p},\theta) = 0$. Thus, we are left with (12) as before (again with the replacement of $\hat{p}$ for $\overline{p}$), and the proposition goes through.

When the constant elasticity form is relaxed, the main result goes through with some modifications. In particular, note that

$$-\left(2x_p + px_{pp}\right) = -x_p\left(2 - x_a x_p B'''\right).$$

It follows that this expression is positive if $B''' < 2ap/[\epsilon(p)x]^2$, where $\epsilon(p)$ is the elasticity of x evaluated at p. Thus, the principle that the effect of water price risk on efficiency depends on the responsiveness of the activity level to the factor price remains unchanged. Further, when water use is highly responsive to the factor price (at least over some range of p), it is more likely that an increase in risk increases factor-use efficiency since it is less likely that the inequality above is satisfied. Within the inelastic range, however, the influence of elasticity is ambiguous.

Before turning to the empirical analysis, we should note that an examination of energy and natural resource markets shows that short-run elasticities of factor utilization can take on a wide range of values. Table 1 displays some estimated own-price elasticities of utilization. The estimated elasticity for residential water is near -0.5, while that for residential electricity is estimated to be roughly twice this figure. Accordingly, the conceptual model developed above implies that we should expect the effect of changes in price risk to vary among these markets. We return to this point in the discussion.

4. EMPIRICAL ANALYSIS

We now discuss a statistical test of the relationship between the diffusion of conservation technology and the magnitude of water price risk. The statistical

Table 1. Elasticities of utilization for selected industries

Resource	*Estimated Elasticity*	*Source*
Residential Electricity	-1.11 to -0.78	Hasset and Metcalf, 1999
Residential Natural Gas	-2.3 to 0.0	Liu, 1982
Gasoline	-1.01 to -0.08	Dahl and Sterner, 1991
Residential Water	-0.64 to -0.46	Nieswiadomy and Cobb, 1993

analysis is facilitated by our use of a unique data set from the water industry. This data set consists of observations on micro-level technology choice in an agricultural water district where farmers are divided into two groups (or "service areas"). Water rates in these service areas are designed such that mean water prices are identical, but prices fluctuate more in one area than in the other. That is, prices are characterized by the mean-preserving spread relationship examined in the last section. Membership in the two service areas is not by choice, but rather is determined by long-standing water rights that are appurtenant to the underlying farm rather than to its owner. Because of the water rights institutions that govern water allocation in the western United States, different price distributions are observed regularly for water users that are otherwise similarly situated (Burness and Quirk, 1979), making water a good candidate for study in this chapter.

The data set consists of observations on water-use technologies at the field level. Having data at this level of disaggregation is critical since environmental conditions such as microclimate, landscape characteristics and soil quality also vary at the field level and are known to exert a large influence on the choice of irrigation technology (Green and Sunding, 1997; Caswell and Zilberman, 1985).

One of our main goals in this section is to test the relationship between the price responsiveness of factor use and the magnitude of price risk. Because the data set is at the field level, we are able to proxy the price responsiveness of water utilization by observing whether the field is dedicated to permanent crops such as trees and vines, or annual crops such as cotton and hay. If the field is planted with a permanent crop, then water use is relatively unresponsive to short-run fluctuations in the price of water. If, however, the farmer produces an annual crop where acreage (i.e., the fraction of the field that is actually planted) fluctuates based on the price, then water use per unit of land is highly responsive to the periodic price of water (Sunding et al., 2002).

4.1 Empirical Model

In the case considered, there are three main water use technologies available. In increasing order of efficiency, these are gravity, high-pressure and low-pressure technologies. Gravity technology includes traditional furrow and flood irrigation systems, high-pressure systems are sprinkler systems, and low-pressure technologies are variants of drip and "microsprinkler" irrigation systems in which water is applied precisely to a plant's root zone. Not surprisingly, gravity systems are the least expensive to design and install and drip systems are the most expensive, with high-pressure systems falling in between.[3]

Although technology choice is discrete, it is possible to order the choice by efficiency to reflect the ranked nature of the alternatives. Let T^* represent the unobserved input-output coefficient of a microunit and assume that it is a linear function of net benefits from investing in technology, that is, $T^* = X\beta + \varepsilon$ where X is a matrix of the explanatory variables, β is a vector of coefficients and ε is the error term, which is assumed to have a standard normal distribution, $\Phi(\varepsilon)$. Let μ_1 and μ_2 represent the cut-off points in the distribution for each possible technology. Technology choice can then be defined in terms of as follows:

$$T = \begin{cases} 0 & \text{if } T^* \leq \mu_1 \\ 1 & \text{if } \mu_1 \leq T^* \leq \mu_2 \\ 2 & \text{if } T^* \geq \mu_2 \end{cases} \tag{16}$$

where $T = 0$ indicates gravity technology is observed, $T = 1$ indicates high pressure technology is observed, and $T = 2$ indicates low pressure technology is observed.

The cut-off points are estimated empirically. In particular, we estimate the following probabilities:

$$Pr(T = 0) = \Phi(\mu_1 - \beta' x) \tag{17}$$
$$Pr(T = 1) = \Phi(\mu_2 - \beta' x) - \Phi(\mu_1 - \beta' x) \tag{18}$$
$$Pr(T = 2) = 1 - \Phi(\mu_2 - \beta' x) \tag{19}$$

Equation (17) provides the structural model for the ordered probit estimation of the adoption of water-use technology. In the following sections we describe the data and estimation results.

4.2 Data

The data used in this analysis is a sample of 1,224 fields served by the Arvin-Edison Water Storage District, located 90 miles north of Los Angeles in California's Central Valley. The data set includes information on water-use technology, environmental conditions, the degree of water price risk and crop choice for a cross-section of 92,294 acres of land observed in 1993. The sample is balanced across available technologies: 44 percent of the fields use gravity irrigation, 21 percent use sprinkler and 35 percent use drip. Tables 2 and 3 provide summary statistics for the data set.

Table 2. Summary statistics for continuous variables

Variable	*Mean*	*Standard Deviation*	*Minimum*	*Maximum*
Permeability (inches/hour)	2.89	3.00	0.13	13
Slope (percent)	1.58	1.32	0.50	10
Field Size (acres)	50.78	52.66	1.00	490

Table 3. Summary statistics for discrete variables

Variable	*Observations*	*Percent of Sample*
Technology Choice		
Gravity	534	43.63
High Pressure	261	21.32
Low Pressure	429	35.05
Crop Choice		
Permanent	960	78.43
Annual	264	21.57
Service Area		
Risk	639	52.21
No Risk	585	47.79

To control for the effect of landscape characteristics on the choice of irrigation technology, we included two environmental variables in our estimation: soil permeability and field slope. Soil permeability is measured in inches per hour and describes how fast the soil drains, or, conversely, how well it retains moisture. In our sample, soil permeability varies from 0.13 inches/hour to 13 inches/hour. Because pressurized irrigation systems can distribute water more evenly over time, these technologies are land-quality augmenting and improve the soil's water storage capacity relative to gravity systems. Thus, we expect soil permeability to have a positive effect on water-use efficiency.

Field slope describes the grade of the field. This variable is measured in percentage terms, where a higher percentage indicates a steeper slope. Slope varies from 0.5 percent to 10 percent in our sample. Since gravity irrigation technologies are difficult to implement on sloped fields, we would expect slope to have a positive effect on optimal efficiency.

The data set also includes the size of each field in acres. Field size can be used to control for scale economies in technology adoption. If there exist scale (dis)economies associated with adoption of efficient irrigation technologies, we would expect the probability of adoption to (decrease) increase with field size. The average field in our sample is 50.8 acres.

The water district in our sample has two service areas. As discussed earlier, water rates in Arvin-Edison are designed so that customers in these service areas face the same mean price of water, but are exposed to different levels of price risk (Arvin Edison Water Storage District, 1982).[4] Owing to the nature of water rights in the western United States, the degree of price risk is a characteristic of the field, and not of its owner. Accordingly, the degree of price risk is denoted as a binary variable (Risk), which is coded as 1 if the field is located in the high-risk service area.

As mentioned earlier, our data set includes information about the type of crop grown on each field. Both annual and permanent crops are evident in the sample: 78 percent of the fields are devoted to permanent crops while the remaining 22 percent are allocated to annual crops. Annual crops grown in Arvin-Edison include primarily lettuce, tomatoes, potatoes and carrots, and permanent crops include oranges, grapes, and tree fruits. Again, we are interested in crop choice primarily as it relates to the effect of price risk on water-use efficiency. The conceptual model predicts that the effect of increasing price risk on the choice of technology depends on the magnitude of the elasticity of utilization. In particular, the model predicts that increasing factor price risk will increase the incentive to adopt efficient technology when the elasticity of utilization is high, and reduce it when the elasticity is low. Thus, risk is included in the model directly and interacted with the crop choice variable.

4.3 Estimation Results

Table 4 presents the parameter estimates for the ordered probit model. As expected, the interaction term between risk and permanent crop production is significant, and the coefficients on the price risk and interaction terms are jointly significant.

This pattern of significance and sign conforms to the predictions of the conceptual model. When farmers produce both annual and permanent crops, the aggregate relationship between price risk and water-use efficiency is ambiguous (increasing risk leads some farmers to increase efficiency and others to decrease efficiency). This argument explains the insignificance of risk alone. The conceptual model does indicate that the influence of risk on efficiency is conditioned on the elasticity of utilization. This observation explains the significance and sign of the interaction term. Taken together, these results provide important confirmation of the theory developed earlier.

Table 4. Ordered probit estimation results

Variable	*Estimated Coefficient*	*Standard Error*	*p-value*
Risk	0.130	0.134	0.333
Risk*Permanent	-0.511***	0.155	0.001
Permanent	0.194*	0.115	0.090
Field Size	0.003***	0.001	0.000
Permeability	0.006	0.012	0.646
Slope	0.395***	0.031	0.000
μ_1	0.565	0.114	-
μ_2	1.189	0.114	-
Test of joint significance of Risk and Risk*Permanent		$\chi^2 = 22.77^{**}$	
LRI (McFadden R^2)			

***Significant at the 1% level,**Significant at the 5% level, *Significant at the 10% level.

In Table 5, we consider how the probability of adopting each of the three types of technologies changes when factor price risk changes.[5] For fields devoted to annual crops, increasing price risk increases the probability of adopting drip irrigation by 5 percent. The effect on sprinkler technologies is negative but small. Looking at the results for fields in permanent crops, we find that an increase in factor price risk decreases the likelihood of adopting drip irrigation by nearly 12 percent. These results suggest that price risk has a large influence on optimal water-use efficiency. The estimated coefficients for field size and slope are also significant.

Table 5. Predicted probability of technology adoption for annual and permanent crops

	Annual Crop			*Permanent Crop*		
Technology	No Risk	Risk	Change*	No Risk	Risk	Change*
Gravity	41.73	36.74	-4.98	36.33	49.14	12.81
High Pressure	24.34	24.46	0.12	24.45	23.49	-0.96
Low Pressure	33.93	38.79	4.86	39.22	27.37	-11.85

*Change in adoption probability resulting from an increase in price risk.

To more easily interpret the ordered probit coefficients, consider the elasticity of the probability of adoption with respect to field size, permeability and slope. Average elasticities are given in Table 6 and are computed as follows. Let j index the technology, n denote the number of observations, and x denote explanatory variables. The average elasticity over all observations is

$$\eta_{xj} = \frac{1}{n} \sum_{n=1}^{1224} \eta_{xn} \; \forall j, x$$

where

$$\eta_{xn} = \frac{\partial \hat{Pr}_{jn}}{\partial x_n} \frac{x_n}{\hat{Pr}_{jn}},$$

and $\hat{Pr}_{jn}$ is the predicted probability of choosing technology j for observation n.

Table 6. Average elasticities for field characteristics

Technology	*Field Size*	*Soil Permeability*	*Slope*
Gravity	-14.6	-1.5	-82.0
High Pressure	-0.9	0.2	-21.8
Low Pressure	13.6	1.8	53.1

A one-percent increase in field size decreases the probability of adopting gravity technologies by 15 percent, increases the probability of adopting high-pressure technologies by 1 percent, and increases the probability of adopting low pressure technologies by 13 percent. Field slope has a large effect on the probability of adoption. A one-percent increase in slope decreases the probability of adopting gravity and high pressure technologies by 82 percent and 22 percent, and increases the probability of adoption low pressure technologies by 53 percent. Soil permeability has a smaller effect, and is statistically insignificant.[6]

5. DISCUSSION

This chapter explores the impact of factor price risk on optimal factor-use efficiency. Since input use efficiency is often embodied in specific capital goods, it is important to consider efficiency in terms of the diffusion of conservation technologies. Toward this end, This chapter first develops a conceptual model of the expected returns from conservation and conventional technologies in an effort to characterize optimal long-run efficiency. An increase in factor price risk is modeled generally using a particular notion of stochastic dominance - the mean-preserving spread. The main conclusion of the conceptual analysis is that the effect of factor price risk on efficiency is conditional on the elasticity of utilization. Accordingly, the influence of price risk on the diffusion of conservation technology should be expected to vary across industries.

The main hypothesis is tested using a unique data set concerning the adoption of water-saving technology. Water is an especially promising industry in which to test the theory developed here because the nature of water rights implies that there can be large, exogenous variations in water price risk among otherwise identical agents. Further, data on technology adoption are readily available (at least for agricultural water use), as are data on the relevant environmental conditions that have a marked effect on the relative productivity of various water-use technologies. Estimation of an ordered probit model produces results that are consistent with the conceptual model. In particular, estimation results are consistent with the main hypothesis that the impact of increasing price risk on input use efficiency depends on the magnitude of elasticity of utilization.

One of the main conclusions of economic research on technology diffusion is that the pattern of adoption over time is explained in large part by differences among potential adopters. This concept, which originates with the work of David, , has implications for our problem. In particular, if potential adopters of conservation technology are heterogeneous, then the impact of price risk on efficiency may be more pronounced at the firm or household level than at the aggregate level. If the input in question is a necessity to some and not to others, then price risk may encourage some users to adopt the conservation technology and others to adopt the conventional one. Thus, economists should consider carefully the impact of price volatility when predicting input-use efficiency.

The results of this chapter also have important implications for public policy. The question of diffusion of conservation technology is only a matter of policy interest when adoption has external benefits. Thus, in situations where price volatility leads to adoption of less efficient input-use technologies, then it may be worthwhile to intervene in the market to stabilize prices. Price risk can be reduced by measures to expand storage capacity, reduce storage losses and improve the conveyance infrastructure. Whether these or similar measures are justified depends on a larger analysis of their costs, and also on a comparison of the welfare costs of stabilization to other policies such as an outright subsidy for adoption.

This chapter also suggests that price risk considerations should be introduced into long-run demand forecasts for key factor markets such as electricity, natural gas, water and the like. Since price risk can have a large effect on factor use efficiency, it may also have a large effect on average factor use over the long run. This chapter also indicates that the effect of price risk on factor-use efficiency should also be accounted for when measuring the demand for factor price insurance, as well as the demand for hedging instruments such as factor price and weather derivatives.

Notes

1. Conservation technologies are not the only type of technology that benefits the environment. Pollution-reducing technologies lower the amount of effluent produced per unit of output. These technologies are not considered in this paper.
2. This argument is essentially an expression of Jensen's Inequatlity
3. See Caswell, 1983 for a detailed description of irrigation technology used in California agriculture.
4. In 1964, the district established an integrated water management plan to mitigate the problem of ground water overdraft and the subsequent environmental damages from overdraft, such as water quality degradation, land subsidence, and eventual aquifer depletion. This plan divided the district into two service areas. The surface water service area consists of irrigators receiving supplies from imported project water through the district's distribution infrastructure. Customers in the ground water service area extract ground water from the underlying aquifer. In wet years, the district stores the excess imported supply in the underlying aquifer, creating a water bank for the surface water service area. In dry years, the banked water is withdrawn to provide reliable supply for customers in the surface service area.
5. Probability of adoption in Table 5 is computed for each service area/crop group at the mean of the landscape variables (50.7 acres, 2.9 inches/hour soil permeability, and 1.6 percent slope).
6. We investigated the exogeneity of crop choice to the model of irrigation technology selection. We could not reject the null hypothesis that crop choice is exogenous using a test of weak exogeneity Smith and Blundell, 1993. Further, we explored whether there was a systematic relationship between crop choice and service area and found that service area by estimating a probit model with crop choice as the dependent variable and service area and other factors as explanatory variables. Service area is not significant in this crop choice model.

References

Abel, A.B., 1983. 'Energy price uncertainty and optimal factor intensity: A mean-variance analysis', *Econometrica*, 51(6), 1839–1846.

Arvin Edison Water Storage District, 1982. The arvin edison water storage district project:a successful water resources management program. Technical report.

Borch, K., 1969. 'A note on uncertainty and indifference curves', *Review of Economic Studies*, 36(105), 1–4.

Boyd, G.A. and S. Karlson, 1993. 'The impact of energy prices on technology choice in the united states steel industry', *Energy Journal*, 14(2), 47–56.

Burness, H.S. and J.P. Quirk, 1979. 'Appropriative water rights and the efficient allocation of resources', *American Economic Review*, 69(1), 25–37.

Carey, J.M. and D. Zilberman, 2002. 'A model of investment under uncertainty: Modern irrigation technology and emerging markets in water', *American Journal of Agricultural Economics*, 84(1), 171–183.

Caswell, M. and D. Zilberman, 1985. 'The choices of irrigation technologies in california', *American Journal of Agricultural Economics*, 67(2), 224–234.

Caswell, M.F., 1983. *The diffusion of low-volume irrigation technology in California agriculture*. PhD thesis, University of California, Berkeley.

Dahl, C. and T. Sterner, 1991. 'Analysing gasoline demand elasticities: A survey', *Energy Econmics*, 13(3), 203–208.

David, P., 1969. Contribution to the theory of diffusion. Memorandum no. 71, Stanford Center for Research in Economic Growth.

Green, G.P. and D.L. Sunding, 1997. 'Land allocation, soil quality, and the demand for irrigation technology', *Journal of Agricultural and Resource Economics*, 22(2), 367–375.

Hardy, G., J. Littlewood and G. Polya, 1934. *Inequalities*, New York: Cambridge University Press.

Hasset, K.A. and G.E. Metcalf, 1999. 'Measuring the energy savings from home improvement investments: Evidence from monthly billing data', *The Review of Economics and Statistics*, 81(31), 516–528.

Hausman, J.A., 1979. 'Individual discount rates and the purchase and utilization of energy-using durables', *The Bell Journal Of Economics*, 10(1), 33–54.

Jaffe, A.B., R.G. Newell and R.N. Stavins, 2001. 'Technological change and the environment' in *Handbook of Environmental Economics*, edited by K-G. Mäler and J. Vincent, Elsevier Science, Amsterdam.

Kneese, A.V., 1978. 'Energy conservation policies', *Natural Resources Journal*, 18(4), 815–823.

Liu, B-C, 1982. 'Natural gas price elasticities: Variations by region and by sector in the USA', *Energy Economics*, 5(3), 195–207.

Nieswiadomy, M. and S.L. Cobb, 1993. 'Impact of pricing structure selectivity on urban water demand', *Contemporary Policy Issues*, 11(3), 101–113.

Pizer, W., W. Harrington, R. Kopp, R. Morgenstern and J. Shih, 2002. Technology adoption and aggregate energy efficiency. Discussion paper 02-52, Resources for the Future.

Rose, N.L. and P.L. Joskow, 1990. 'The diffusion of new technologies: Evidence from the electric utility industry', *RAND Journal of Economics*, 21(3), 354–373.

Rothschild, M. and J.E. Stiglitz, 1970. 'Increasing risk: 1. a definition', *Journal of Economic Theory*, 2(3), 225–243.

Smith, R.J. and R.W. Blundell, 1993. 'An exogeneity test for a simultaneous equation tobit model with an application to labor supply', *Econometrica*, 11(3), 101–113.

Sunding, D., D. Zilberman, R. Howitt, A. Dinar and N. MacDougall, 2002. 'Measuring the costs of reallocating water from agriculture: A multi-model approach', *Natural Resources Modeling*, 15(2), 201–225.

Sunding, D. and D. Zilberman, 2001. 'The agricultural innovation process: Research and technology adoption in a changing agricultural sector' in *Handbook of Agricultural Economics*, edited by B.L. Gardner and G.C. Rausser, Elsevier Science, Amsterdam.

Chapter 6

OPTIMAL MANAGEMENT OF GROUNDWATER OVER SPACE AND TIME

Nicholas Brozović
University of Illinois at Urbana-Champaign, U.S.A.

David Sunding
University of California, Berkeley, U.S.A.

David Zilberman
University of California, Berkeley, U.S.A.

1. INTRODUCTION

For nearly half a century, groundwater has been portrayed in the economic literature as a typical common property resource. Numerous studies of groundwater extraction have analyzed the externalities imposed by users on each other.[1] A large body of work offers clear prescriptions in the form of optimal policy instruments, and a similarly large body of work advocates the needlessness of any centralized intervention. Yet existing theoretical models of groundwater extraction implicitly make two strong assumptions about the underlying behavior of the resource. First, the spatial distribution of resource users is assumed to be irrelevant. Second, path-independence of the resource is assumed: the history of past extraction does not affect present and future extraction decisions. Relaxing either of these assumptions may undermine the results of existing work.

The purpose of this chapter is to present a model for the extraction of a path-dependent resource by spatially distributed users. The example of groundwater is used to demonstrate the incorporation of the physics of a complex natural system into an economic model of dynamic resource use. In particular, the optimality conditions can be calibrated to parameters found in actual aquifers to model the range of behavior encountered in the real world. This demonstrates the failure of existing models of groundwater extraction to describe aquifers adequately.

The analysis presented in this chapter emphasizes the tradeoffs between the spatial extent of each user's private property right, the physical parameters of the system, and the spatial and temporal distribution of extraction. Several important principles emerge from the model. Some aquifers, even if they constitute a single hydrological entity: (1) are more akin to private property than common property (see the end of section 3.2), and (2) have significant lagged effects from pumping (see Proposition 2). In such cases, use of traditional dynamic common property models will result in misleading or incorrect analyses and policy prescriptions. The model presented is quite general and can also be applied to other resources where externalities are diffusional in nature, such as oilfields or patchy marine fisheries.

This chapter is organized into several sections. We begin with a simple description of the physics of groundwater flow and contrast this to the representation of flow in existing economic models of groundwater. Following this, a theory for the optimal extraction of groundwater by multiple spatially distributed users from a hydrologically realistic, path-dependent aquifer is presented. Although the model we present is intended to allow incorporation of groundwater flow equations taken from the engineering and hydrology literature, it is also general enough to nest many existing economic models of groundwater use (see Appendix C). Discussion of the optimality conditions from this model emphasizes how the results differ from existing studies and the implications for groundwater management policy.

2. A SIMPLE DESCRIPTION OF THE HYDRAULICS OF GROUNDWATER FLOW

Ongoing pumping from a well in an aquifer induces horizontal hydraulic gradients towards the well. Because of these gradients, a localized 'cone of depression' develops around the well. The dimensions of a cone of depression will depend not only on the pumping rate through time, but also on the hydrogeological variables that describe the physical properties of the aquifer (see Appendix A). However, for an aquifer with homogeneous physical properties, a well pumping large quantities of water will have a deeper, wider cone of depression than a well pumping small quantities.

Moreover, if the cones of depression of adjacent wells overlap, well interference will occur and the water level in both wells will decrease correspondingly. Because of the physics of water flow, even though well interference is spatially

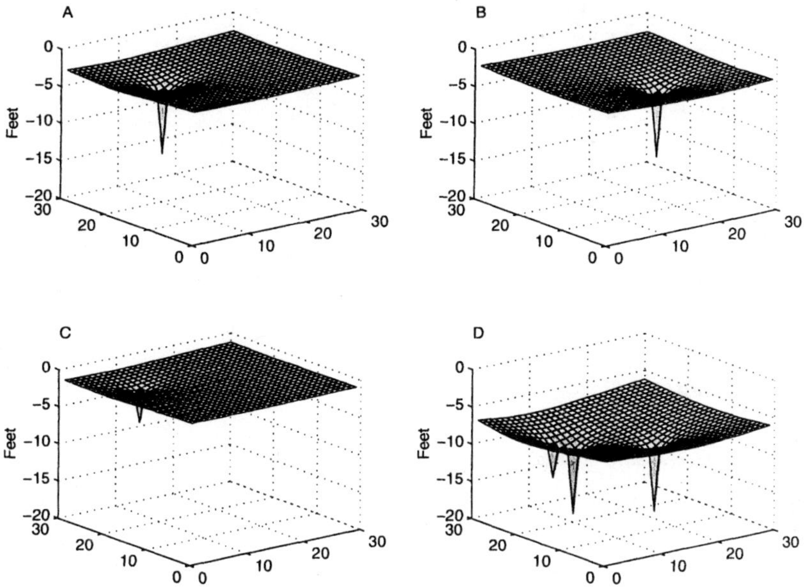

Figure 1. Drawdown from multiple wells in an aquifer. Panels A through C represent the drawdown caused by three separate wells in an aquifer whose behavior is governed by the hydraulic response equations (see Appendix A). In each panel, the units of the vertical axis are feet, and of the horizontal axes, miles. Before the start of pumping, the aquifer was assumed to have a uniform depth of zero feet. Panels A and B show the spatial distribution of drawdown after one year resulting from two wells in different locations, each pumping 600,000 gallons a day. Panel C shows the distribution of drawdown after a year for a third well pumping 300,000 gallons a day. Panel D shows the resultant drawdown if all three wells pumped simultaneously for a year. Storativity and transmissivity values are within the range found in normal aquifers (storativity taken as 10^{-3}, transmissivity as 10^5 gal/day/ft). Note that (1) cones of depression are localized to the vicinity of each pumping well; (2) drawdown is greater for the wells with larger pumping rates; and (3) well interference is greater between wells that are closer together (see Panel D).

variable, it is also linearly additive. Hence, the total drawdown of the aquifer at any point caused by pumping from any number of wells is the sum of the drawdowns caused by each individual well at that point (Figure 1). Aquifers show an important additional behavior in response to withdrawals of water. As described in Appendix A, the water level in a well does not adjust instantaneously to changes in pumping rate. Instead, adjustments to changes in pumping are gradual and cumulative. Thus, the entire history of water extractions determines the state of the groundwater resource at any point in time.

In most real-world aquifers, there are multiple independent and heterogeneous users that each pump groundwater. Each user's pumping will affect the pumping costs of all the other users. Each possible pair of users will thus have an idiosyncratic set of effects on each other. Bilateral impacts will depend on both the distance between the two users and the history of past pumping at each well. Moreover, these impacts will be lagged: a change in one user's behavior may not be observed by other users for some time.

Economic studies of groundwater extraction use one of three different models to represent aquifers: single-cell, two-cell, and multi-cell. None of these models adequately capture either the spatial interdependency among pumpers or the path-dependency property described above. However, in order to understand exactly how the model described in this chapter differs from previous work, each type of model and its implicit assumptions will be discussed.

The simplest aquifer representation is the single-cell aquifer (first described in detail by Brown and Deacon, 1972). In a single-cell aquifer, the state of the groundwater resource is entirely described by a single variable, generally the volume of water remaining in the aquifer or the depth to water. This aggregation of the resource stock represents an implicit assumption that the water level is uniform throughout the aquifer. Because of this, single-cell models are often referred to as 'bathtub' or 'milk-carton' models. In such a system, no matter where, or from how many places in the bathtub (or milk-carton) liquid is extracted, the depth of the liquid throughout the container remains uniform. Hence, in an unconfined single-cell aquifer, drawdown of the water table is uniform throughout the aquifer irrespective of both the location of pumping wells relative to each other and their relative contributions to the aggregate extraction (Figure 2). Although in principle, path-dependency of the resource could be incorporated into a single-cell models, to date this has not been undertaken. Instead, in discrete-time formulations, changes in the resource depend only on the previous period's extraction (Burt, 1970; Feinerman, 1988). In continuous time formulations, the resource stock adjusts instantaneously to the extraction rate (for example, Gisser, 1983; or Koundouri, 2004). The focus of this chapter is the presentation and analysis of a path-dependent groundwater extraction model with spatial heterogeneity. However, for ease of comparison, Appendix B derives the optimality conditions for groundwater extraction from a single-cell aquifer in discrete time. Appendix C shows how this single-cell model nests within our more general framework.

A somewhat more complicated aquifer representation is the two-cell model, where several single-cells are mutually connected by porous boundaries (Chakravorty and Umetsu, 2003; Eswaran and Lewis, 1984; Khalatbari, 1977; Zeitouni and Dinar, 1997).[2] Each component cell in a two-cell model behaves exactly like a single-cell. There is also flow between the two cells that is proportional to the difference in stock levels between them. However, in existing models the

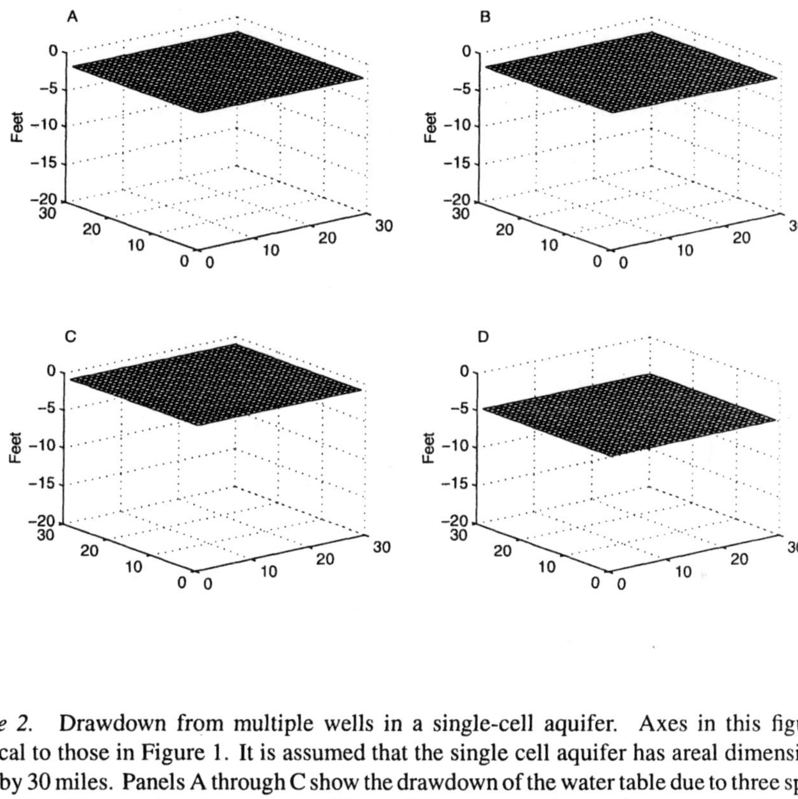

Figure 2. Drawdown from multiple wells in a single-cell aquifer. Axes in this figure are identical to those in Figure 1. It is assumed that the single cell aquifer has areal dimensions 30 miles by 30 miles. Panels A through C show the drawdown of the water table due to three spatially separated wells pumping for one year. The well locations and pumping rates are the same as in Figure 1. Panel D shows the resultant drawdown if all three wells pumped simultaneously for a year. Note that Panels A and B are identical, and that the depth to the water table remains uniform, irrespective of the position of pumping.

rate of this adjustment only depends on instantaneous stock differences between the component cells, and there is no role for extraction history.

Finally, in multi-cell aquifer models, water movement between cells is determined by finite difference approximations to the equations of groundwater flow (Bredehoeft and Young, 1970; Noel et al., 1980; Noel and Howitt, 1982). Multi-cell models are usually calibrated to individual groundwater basins and provide specific management guidelines rather than general results. Early contributions to this literature did not involve any optimization, but instead compared the effects of simple rule-of-thumb policies (Bredehoeft and Young, 1970). More recent work has used separate physical models of groundwater behavior and economic models of the benefits of water use. Most of these papers simulate aquifer behavior under various pumping scenarios and then use a linear regression of the physical model as the vector state equation in the economic modeling (Noel et al., 1980; Noel and Howitt, 1982). The numerical simulations employed in such models are generally hydrologically accurate. As such, they do capture path-dependency of the groundwater resource. However, in linearizing the physical model for inclusion in the economic model, only the previous period's state and control variables are used. This is a misspecification of the physical model that removes the role of extraction history. Hence, lagged groundwater pumping externalities cannot be analyzed in an economic context using such models.

3. OPTIMAL EXTRACTION OF A PATH-DEPENDENT RESOURCE BY SPATIALLY DISTRIBUTED USERS

3.1 The model

Consider an aquifer whose behavior is governed by the hydraulic response equations (A. 4) to (A. 6) described in Appendix A to this chapter. Water is to be extracted from the aquifer by J separate users over an N-period horizon. These users are spatially distributed with known, fixed locations relative to each other and to the resource, and each owns a single well. [3] In any period $t = 1, \ldots, N$, the net benefit of each user $j = 1, \ldots, J$ from the resource is given by the function $f(u_{jt}, x_{jt})$, which captures both the benefits and costs of resource extraction. The decision variable u_{jt} is user j's per-period water extraction at time t. Assume that $f(u_{jt}, x_{jt}) > 0$, $\partial f(u_{jt}, x_{jt})/\partial u_{jt} > 0$ and $\partial^2 f(u_{jt}, x_{jt})/\partial u_{jt}^2 < 0$. The state variable x_{jt} is defined as the pumping lift of water at the jth well at time t. Assume that $\partial f(u_{jt}, x_{jt})/\partial x_{jt} < 0$, as per-period benefits decrease as the pumping lift increases. Also assume that $\partial^2 f(u_{jt}, x_{jt})/\partial u_{jt}^2 \leq 0$, so that pumping costs increase at least linearly with depth. Note that in every period, x_{jt} is determined not only by user j's previous extraction history, but also by the extraction history of all the other users. Appendix A provides more details on the determination of pumping lifts through time. The variable $s(t, r)$ in that appendix is equivalent to x_{jt} in the model presented in this section; here we follow standard optimization notation for state and control variables for ease of interpretation. Similarly, the difference in pumping rates for a well between successive periods, $(u_{jt} - u_{jt-1})$, is equivalent to the incremental pumping ΔQ in Appendix A.

The N-period optimization problem for the aquifer is given by

$$\max \sum_{t=1}^{N} \beta^t \sum_{j=1}^{J} f(u_{jt}, x_{jt}) \tag{1}$$

where β is the per-period discount factor, with $\beta < 1$. The aquifer is spatially heterogeneous, so that each well will have a different pumping lift determined by all previous pumping histories. The J equations of motion describing the level of water over time in each of the J wells are

$$x_{jt+1} = \sum_{i=1}^{J} \sum_{n=1}^{t} (u_{in} - u_{in-1}) w(t - n + 1, r(i, j)); \; t = 1, \ldots, N-1 \tag{2}$$

where $w(t, r(i, j))$ is the well function defined by equation (A.4) and $r(i, j)$ is the distance between wells i and j. Note that there are no 'cells' in this analysis. The pumping lifts in the aquifer resulting from drawdown of multiple wells form a continuous surface and are defined for every point in the aquifer. However, only the lifts at pumping wells enter the objective function. Without loss of generality, it is assumed that $x_{j0} = 0$ for all j. Note that because the state of the resource at all periods after N is unimportant, there are only $(N-1)J$ equations of motion in total.

Whereas in many problems in the optimal extraction of resources over time, the Hamiltonian and optimal control theory are the most convenient solution concepts, this is not the case here. In the discrete time formulation, the presence of lagged effects leads to equations of motion for the resource that are summations rather than difference equations. Because of this, the method of Lagrange multipliers is more convenient in order to derive the necessary conditions for this problem. [4] The appropriate Lagrangian expression for the problem described by equations (1) and (2) is

$$\begin{aligned} L = & \sum_{t=1}^{N} \beta^t \sum_{j=1}^{J} f(u_{jt}, x_{jt}) + \\ & + \sum_{t=1}^{N} \sum_{j=1}^{J} \lambda_{jt} \left\{ \left(\sum_{i=1}^{J} \sum_{n=1}^{t} (u_{in} - u_{in-1}) w(t - n + 1, r(i, j)) \right) - x_{jt+1} \right\} \end{aligned} \tag{3}$$

The first order conditions for an interior solution are:

$$\frac{\partial L}{\partial x_{ls}} = \beta^s \frac{\partial f(u_{ls}, x_{ls})}{\partial x_{ls}} - \lambda_{ls-1} = 0 \tag{4}$$

$$\begin{aligned}\frac{\partial L}{\partial u_{ls}} &= \beta^s \frac{\partial f(u_{ls}, x_{ls})}{\partial u_{ls}} + \sum_{j=1}^{J} \lambda_{js} w(1, r(l,j)) + \\ &+ \sum_{t=s+1}^{N} \sum_{j=1}^{J} \lambda_{jt} \left(w(t-s+1, r(l,j)) - w(t-s, r(l,j)) \right) = 0\end{aligned} \tag{5}$$

By definition, $w(0, r(l,j)) = 0$, so that condition (5) may be rewritten in more compact form as

$$\begin{aligned}\frac{\partial L}{\partial u_{ls}} &= \beta^s \frac{\partial f(u_{ls}, x_{ls})}{\partial u_{ls}} + \\ &+ \sum_{t=s}^{N} \sum_{j=1}^{J} \lambda_{jt} \left(w(t-s+1, r(l,j)) - w(t-s, r(l,j)) \right) = 0\end{aligned} \tag{6}$$

The adjoint variable λ_{jt} is the marginal present value shadow price of the state variable at well j at time t. Defining the transformation $\lambda_{jk} = \beta^k \mu_{jk}$ where μ_{jk} is the marginal current value shadow price of water[5] at well j at time k allows us to restate conditions (4) and (6) in current value form:

$$\frac{\partial f(u_{ls}, x_{ls})}{\partial x_{ls}} - \beta^{-1} \mu_{ls-1} = 0 \tag{7}$$

$$\frac{\partial f(u_{ls}, x_{ls})}{\partial u_{ls}} + \sum_{t=s}^{N} \sum_{j=1}^{J} \beta^{t-s} \mu_{jt} \left(w(t-s+1, r(l,j)) - w(t-s, r(l,j)) \right) = 0 \tag{8}$$

The double summation in condition (8) may be written in simplified notation as

$$\frac{\partial f(u_{ls}, x_{ls})}{\partial u_{ls}} + \sum_{t=s}^{N} \sum_{j=1}^{J} \beta^{t-s} \mu_{jt} \theta(t-s, r(l,j)) = 0 \tag{9}$$

The function $\theta(t-s, r(l,j))$, which is the difference between well functions in successive time periods, is the incremental drawdown caused at well j at time t by a unit of pumping at well l at time s. Sufficient conditions for optimality are joint concavity of $f(u_{ls}, x_{ls})$ in u_{ls} and x_{ls}.

3.2 Results

Equation (7) shows that for an optimal solution, the marginal benefit to each groundwater user of a further unit of pumping lift equals the difference between the capital gain and opportunity costs to that user of the additional pumping lift. Equation (9) relates the benefit of pumping an additional unit of water to the discounted future costs of that pumping for all users. Hence, condition (9) captures the lagged, idiosyncratic effects of resource extraction. Several key insights about the behavior of the optimal solution emerge from these two necessary conditions.

PROPOSITION 1 *(Role of spatial interdependency): The further a well is from its nearest neighbor wells, the larger its optimal pumping in each period.*

Proof: Because summation is a linear operator, we can demonstrate the result using only two wells without loss of generality. Recall that $f(u_{jt}, x_{jt}) > 0$, $\partial f(u_{jt}, x_{jt})/\partial u_{jt} > 0$ and $\partial^2 f(u_{jt}, x_{jt})/\partial u_{jt}^2 < 0$. Moreover, the adjoint variable is negative by definition of the state variable. Hence, from equation (9), we need to show that

1 $(w(t+1,r) - w(t,r)) > 0$ and

2 $\partial/\partial r \, \{w(t+1,r) - w(t,r)\} < 0$.

The first result follows immediately from the definition of the well function in equation (A. 4), as

$$\begin{aligned} w(t+1,r) - w(t,r) &= \frac{1}{4\pi T}\left(\int_{\frac{r^2S}{4T(t+1)}}^{\infty} \frac{e^{-z}}{z} dz - \int_{\frac{r^2S}{4Tt}}^{\infty} \frac{e^{-z}}{z} dz\right) \\ &= \frac{1}{4\pi T}\left(\int_{\frac{r^2S}{4T(t+1)}}^{\frac{r^2S}{4Tt}} \frac{e^{-z}}{z} dz\right) > 0 \qquad (10) \end{aligned}$$

To show the second result, note that

$$\begin{aligned} \frac{\partial}{\partial r}\{w(t+1,r) - w(t,r)\} &= \frac{1}{4\pi T}\left(\frac{e^{\frac{-r^2S}{4Tt}}}{r^2S/4Tt} \cdot \frac{2rS}{4Tt} - \right. \\ &\qquad \left. - \frac{e^{\frac{-r^2S}{4T(t+1)}}}{r^2S/4T(t+1)} \cdot \frac{2rS}{4T(t+1)}\right) \\ &= \frac{1}{2\pi Tr}\left(e^{\frac{-r^2S}{4Tt}} - e^{\frac{-r^2S}{4T(t+1)}}\right) \\ &< 0 \qquad (11) \end{aligned}$$

Thus, because the magnitude of the externality imposed by one user on another depends on their distance from one another, where two users are close together, they will each optimally pump less water. ||

PROPOSITION 2 *(Role of extraction history): The maximum effect of a user's pumping need not be felt immediately. As distance from a pumping well increases, the time lag between a change in pumping at that well and the maximum effect of that pumping will also increase.*

Proof: Two separate results are needed:

1 The sign of $\partial/\partial t\,\{w(t+1,r) - w(t,r)\}$ is ambiguous. This implies that the effects of a given change in pumping as felt at any distance r may increase or decrease with time.

2 As r increases, the time $\widehat{t}$ at which $\partial/\partial t\left\{w(\widehat{t}+1,r) - w(\widehat{t},r)\right\} = 0$ also increases.

To show the first result, calculate the appropriative derivative:

$$\begin{aligned}\frac{\partial}{\partial t}\{w(t+1,r) - w(t,r)\} &= \frac{1}{4\pi T}\left(\frac{e^{\frac{-r^2S}{4T(t+1)}}}{t+1} - \frac{e^{\frac{-r^2S}{4Tt}}}{t}\right) \\ &= \frac{1}{4\pi T(t+1)}\left(e^{\frac{-r^2S}{4T(t+1)}} - \frac{t+1}{t}e^{\frac{-r^2S}{4Tt}}\right) \\ &= \frac{e^{\frac{-r^2S}{4Tt}}}{4\pi T(t+1)}\left(e^{\frac{r^2S}{4Tt(t+1)}} - \frac{t+1}{t}\right) \qquad (12)\end{aligned}$$

Now, note that for large t, $e^{\frac{r^2S}{4Tt(t+1)}} - \frac{t+1}{t}$ will be negative, whereas for large r and S, and small t, it will be positive. Thus, the sign of $\partial/\partial t\{w(t+1,r)$ $-w(t,r)\}$ is ambiguous.

To show the second result, define $\widehat{t}$ such that $e^{\frac{r^2S}{4T\widehat{t}(\widehat{t}+1)}} - \frac{\widehat{t}+1}{\widehat{t}}$, so that $\partial/\partial t\{w(\widehat{t}+1,r) - w(\widehat{t},r)\} = 0$. From this it is clear that if r increases, $\widehat{t}$ must also increase. ||

PROPOSITION 3: *A spatially uniform policy will only be optimal if there are an infinite number of wells uniformly distributed above the aquifer.*

Proof: In order for any uniform policy to be optimal, the double summation $\sum_{t=s}^{N}\sum_{j=1}^{J}\beta^{t-s}\mu_{jt}\theta(t-s, r(l,j))$ must be equal for all pairs of well users j and l, and for all periods s. From Propositions 1 and 2, this can only be true if every well has the same spatial distribution of wells around it. ||

From Proposition 3, it follows that if there are an infinite number of uniformly distributed wells, the optimal policy will be spatially uniform. However, unlike in a single-cell aquifer, this does not mean that the resource is common property. Idiosyncratic externalities are still present, but each well receives the same overall distribution of idiosyncratic effects. Hence, the more spatially non-uniform well distribution is, the more the optimal policy will also be non-uniform across space, even if the individual resource users have identical net benefit functions for water, as is the case in this analysis.

Assuming that the initial condition of the aquifer is not at the optimal steady state, and that $N = \infty$, equations (7) and (9) allow solution of the optimal trajectory to reach that steady state. Solution of this system of equations is computationally intensive, and requires explicit spatial locations for each groundwater user. However, analysis of the optimal steady state is also informative. Given the assumptions made about the infinite areal extent of the aquifer (see Appendix A), every finite pumping combination will reach a steady state.[6] The optimal steady state is defined by a set of state variables $x_1^*, x_2^*, \ldots, x_J^*$ and a set of control variables $u_1^*, u_2^*, \ldots, u_J^*$. In a steady state, condition (7) implies that

$$\mu_l^* = \beta \frac{\partial f(u_l^*, x_l^*)}{\partial x_l^*} \tag{13}$$

Substituting into condition (9) yields

$$\frac{\partial f(u_l^*, x_l^*)}{\partial u_l^*} + \sum_{j=1}^{J} \frac{\partial f(u_j^*, x_j^*)}{\partial x_j^*} \sum_{t=s}^{\infty} \beta^{t-s+1} \theta(t-s, r(l,j)) = 0 \tag{14}$$

Now, the infinite series of well functions in the second term of the left hand side is a convergent sequence with finite sum, so that the steady state condition may be simplified to

$$\frac{\partial f(u_l^*, x_l^*)}{\partial u_l^*} = -\sum_{j=1}^{J} \frac{\partial f(u_j^*, x_j^*)}{\partial x_j^*} \sum_{t=1}^{\infty} \beta^t \theta(t-1, r(l,j)) \tag{15}$$

Equation (15) relates the optimal steady state marginal value of pumping to the discounted cost to all users of that additional unit of pumping in the future. The summation $\sum_{t=1}^{\infty} \beta^t \theta(t-1, r(l,j))$ can be thought of as a weighting function that determines the relative importance placed on each user's steady state marginal benefit by user l. It captures both the spatial interdependency between pairs of users and the lagged nature of the groundwater externality.

For an aquifer system with two groundwater users, condition (15) may be represented in a convenient graphical form. Although such as a system only contains a single bilateral relationship, linearity of the summation operator means that key features of the optimality condition are preserved. Moreover, such a graphical analysis provides an intuitive way to demonstrate the important differences between this model and existing groundwater economics models.

First, consider the discounted components of the weighting function from equation (15), which for each period t are given by $\beta^t\theta(t-1, r(l,j))$. For a given distance between the two wells j and l, and for a given set of hydrological parameters S and T, a plot of $\beta^t\theta(t-1, r(l,j))$ against time shows the importance of lagged effects in determining when the effects of pumping by one user (namely drawdown of water in the well) are transmitted to the other user (Figures 3 and 4).

As expected from Proposition 2, if the two pumping wells are a small distance apart, only pumping in the immediate past has any relevance. The majority of the impact resulting from any change in pumping occurs immediately (Figure 3). No significant additional drawdown occurs more than several periods after a change in the pumping schedule.

Recall that from Proposition 2, lagged effects become more and more important as distance between the users increases. Hence, at larger distances from a pumping well, users feel no immediate effects from changes in the other users' pumping. Instead, the effects of changes that occurred several periods ago are much more significant (Figure 4). Indeed, even with discounting, users a large distance apart from each other place much more importance on the other's actions many periods ago, and no weight on their present actions. Moreover, the impacts of changes in pumping may persist for many years. Note also the difference in the magnitude of the per-period weighting function between Figures 3 and 4. At a distance of around 10 miles from a pumping well, the second user is far less concerned with changes in the other user's pumping than at a distance of 1000 feet.

At an optimal steady state, the summation $\sum_{t=1}^{\infty}\beta^t\theta(t-1, r(l,j))$ represents the time-integrated total importance to a groundwater user of a unit change in pumping by any other user. If we assume an effective well radius for each user, this weighting function is also defined for the future effects of a user on the water levels in his own well. By normalizing the weighting function by a user's own weighting function, it is possible to consider the relative importance that a user places on other user's groundwater withdrawals as a function of distance (Figures 5 and 6). By definition, a user will place a relative value of one on withdrawals from his own well. A bilateral relationship with a neighboring well that has a relative value of 0.9 implies that the user cares almost as much about withdrawals from this well as about his own withdrawals. Conversely, a value of 0.1 suggests that the two wells interfere very little with each other.

As might be expected, in aquifers with high storativities and low transmissivities, the relative weighting functions decrease rapidly with distance (Figure

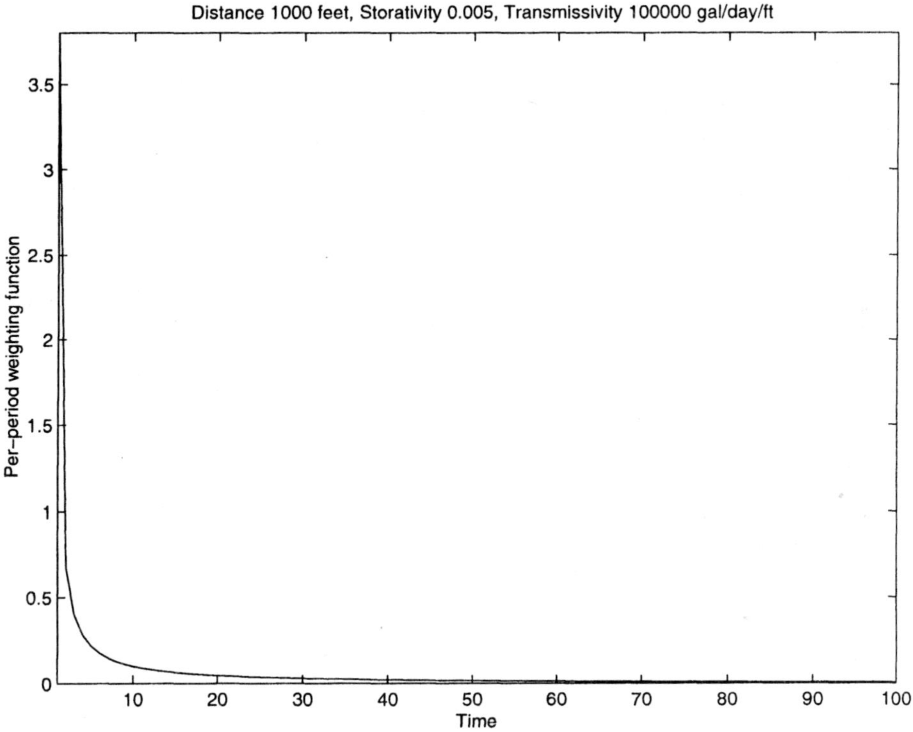

Figure 3. Per-period weighting function through time for a steady-state aquifer. This graph presents the impacts through time, per unit time, of a unit change in pumping from a well on a second groundwater user located 1000 feet from the first well. The units of time are in months. Storativity and transmissivity values are within the range found in normal aquifers (storativity taken as 10^{-3}, transmissivity as 10^5 gal/day/ft). Note that the per-period weighting function $\beta^t\theta(t-1, r(l,j))$ is discounted.

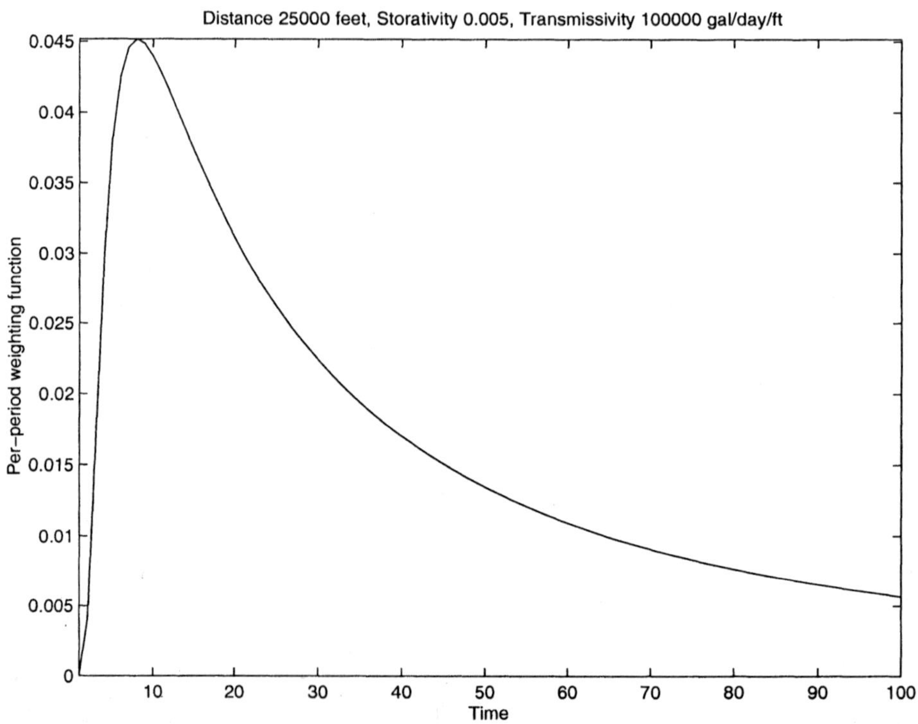

Figure 4. Per-period weighting function through time for a steady-state aquifer. This graph presents the impacts through time, per unit time, of a unit change in pumping from a well on a second groundwater user located 50000 feet from the first well. The units of time are in months. Storativity and transmissivity values are within the range found in normal aquifers (storativity taken as 10^{-3}, transmissivity as 10^5 gal/day/ft). Note that the per-period weighting function $\beta^t\theta(t-1, r(l,j))$ is discounted.

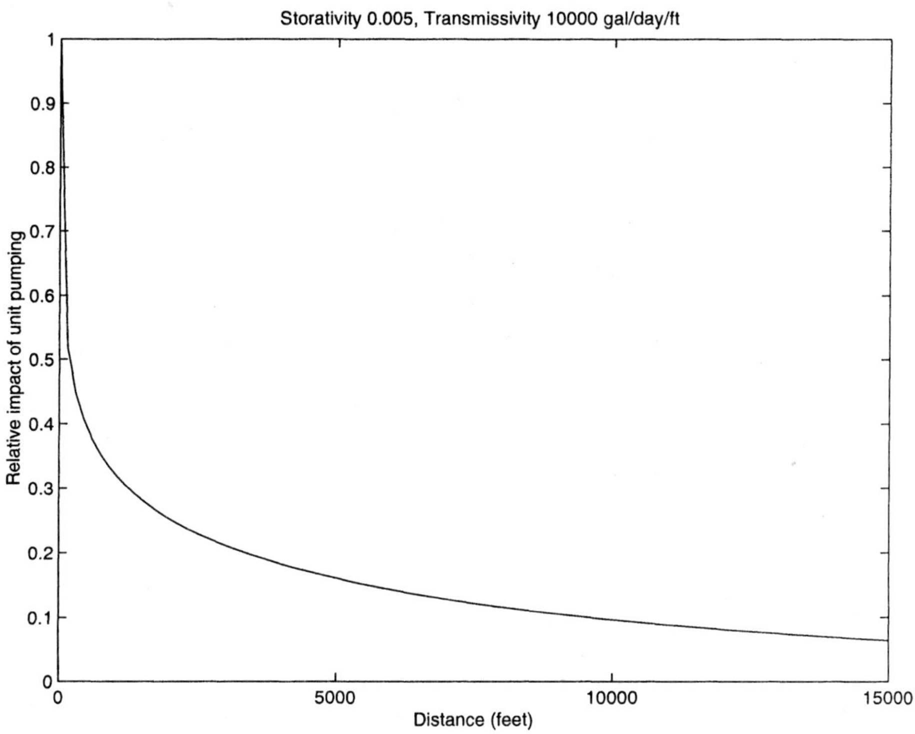

Figure 5. Total relative impacts of pumping as a function of distance. The vertical axis is the normalized weighting function, defined for a distance r as $\frac{\sum_{t-1}^{\infty} \beta^t \theta(t-1,r)}{\sum_{t-1}^{\infty} \beta^t \theta(t-1,1.5)}$. The normalized weighting function is the total weighting function at r divided by the weighting function measured at the effective well radius, taken here as 1.5 feet. Figure 5 represents an aquifer with high storativity and low transmissivity. The graph can be interpreted as follows. A unit of water withdrawn by user j one mile away from user l will have less than 20% of the impact that user l will have on himself through withdrawing one unit of water. Similarly, the transmitted effect for a pumping well at a distance of 3 miles is less than 10% of the own-effect.

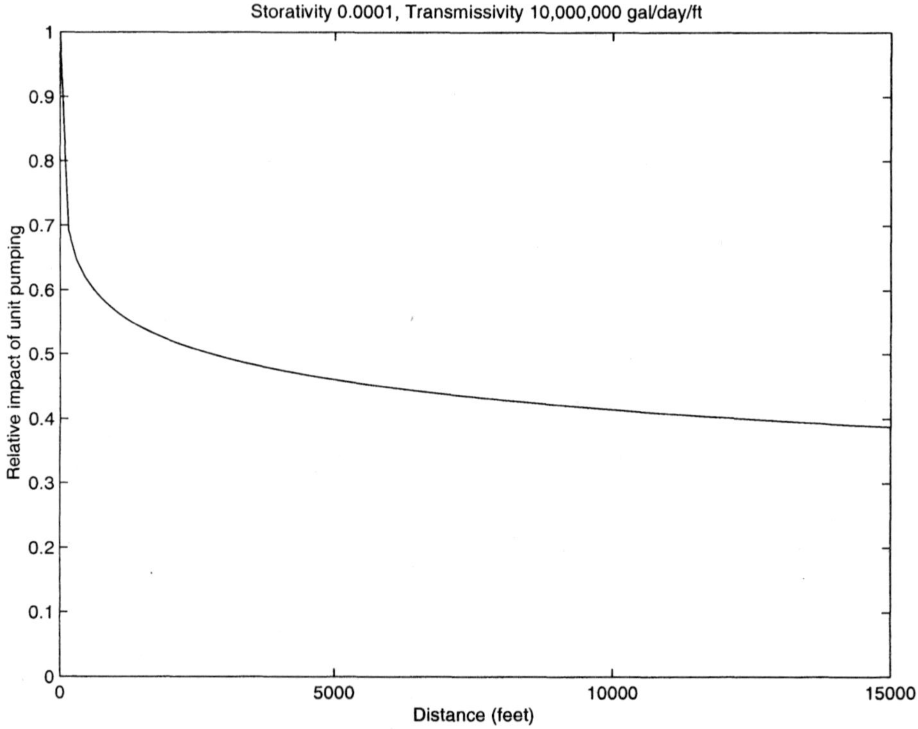

Figure 6. Total relative impacts of pumping as a function of distance. The vertical axis is the normalized weighting function, defined for a distance r as $\frac{\sum_{t-1}^{\infty} \beta^t \theta(t-1,r)}{\sum_{t-1}^{\infty} \beta^t \theta(t-1,1.5)}$. The normalized weighting function is the total weighting function at r divided by the weighting function measured at the effective well radius, taken here as 1.5 feet. Figure 6 represents an aquifer with low storativity and high transmissivity. The graph can be interpreted as follows. A unit of water withdrawn by user j one mile away from user l will have around 45% of the impact that user l will have on himself through withdrawing one unit of water. Similarly, the transmitted effect for a pumping well at a distance of 3 miles is still almost 40% of the own-effect.

5). This implies that in general, groundwater users are unconcerned about other users' extraction rates at any distances away from their wells. As a result of this, we suggest that some aquifers with very high storativities and very low transmissivities should not be modeled as common property. On the other hand, in aquifers with low storativities and high transmissivities, the values of the relative weighting function remain high even at large distances (Figure 6). In such aquifers, each groundwater user's extraction does impact all other users. However, note that for all realistic hydrogeological parameter ranges, the greatest impact on the water level in any well is always caused by pumping from that well. By comparison, in single-cell aquifer models, the relative weighting function is one for all groundwater users, irrespective of distance from one another. This is another way of stating the implicit assumption of single-cell models that extraction from any well affects all users equally (to see this graphically, compare Figures 1 and 2). As Figures 5 and 6 suggest, this assumption may be quite unrealistic.

4. POLICY IMPLICATIONS

An influential body of literature has focused on the magnitude of the welfare gain from optimal control of groundwater compared to competitive outcomes (Allen and Gisser, 1984; Gisser, 1983; Gisser and Sanchez, 1980; Koundouri, 2004). This work has emphasized the apparently negligible welfare difference between optimal control rules and competitive outcomes without any government intervention.

Whether advocating optimal management or no intervention, all of these studies have used single-cell aquifers. As explained in this paper, such models fail to capture adequately important aspects of the behavior of real aquifers. Because of this, policy recommendations based on such models, even when they provide both apparently robust and intuitively appealing results, should be viewed with caution.

As the analysis above has demonstrated, many groundwater aquifers should not be modeled as common property. Under certain hydrological conditions (such as shown in Figure 6), effects of pumping may be widely transmitted throughout the aquifer. However, in other aquifers, the extent of the externality imposed by one user on other users is limited (Figure 5). In such settings, the aquifer is more akin to a private property resource than a common property resource.

Herein lies the failure of the single-cell model to capture adequately aquifer hydraulic response. Single-cell models predict that there are few gains to be made from optimal groundwater management in aquifers with high storativities (Gisser, 1983). With the more realistic aquifer response function, it remains true that there are the least gains from optimal groundwater management in high storativity aquifers. However, such aquifers also least resemble a single-cell as the aquifer response is localized to the immediate vicinity of the pumping wells. In other words, gains from optimal management are minimized because the resource is close to private property to begin with. Conversely, for the situation which most resembles a single-cell aquifer (low storativity and high transmissivity), the gains from optimal groundwater management over no intervention will be larger.

Given the complexity of the underlying resource, it is not surprising that the optimal policy should vary idiosyncratically across space and time. Clearly, it is not feasible to implement such a policy in real groundwater management situations. Which second-best instrument will have the best equity and efficiency effects will depend on the spatial distribution of wells, as well as the individual demand functions for water and local hydrological parameters. It is not possible to rank second-best policies without extensive numerical simulations.

It is a common preconception that in the United States, there is an almost complete lack of groundwater regulations. Whether advocating the introduction of new policies or the needlessness of any intervention, this notion has underlain many of the economic studies of groundwater. However, it is not correct to say that groundwater regulations are generally absent. Many states have well-spacing regulations that determine the minimum distance between adjacent wells. Moreover, there is a large variation in these regulations, from well-spacing requirements of 4 miles in portions of the Dakota aquifer in Kansas, to 300 feet or less in many counties in Texas.

Well-spacing regulations cannot be analyzed at all using a single-cell aquifer model. Because of this, even though they represent a pervasive environmental regulation, they have been entirely ignored in the economic literature. Indeed, because this study shows that the greatest impacts from any pumping are always closest to the well head, it is likely that a simple well-spacing regulation will have excellent efficiency and equity effects in some aquifers. It is conceivable that under some conditions, well-spacing regulations are more appropriate than uniform taxes or quotas. The extent to which actual well-spacing regulations reflect underlying hydrological parameters, and how they correspond to an economically defined optimal spacing, are empirical questions left to future work.

Like water, oil is a fugitive resource, and the same equations of flow govern its subsurface behavior. However, in the oil industry, there has been a widespread failure of well-spacing regulations to prevent over-exploitation (Libecap and Wiggins, 1985; Wiggins and Libecap, 1984). One possible explanation of this is that the extraction rate of oil implies well-spacing regulations that are impossible to enforce given the surface area of individual oil leases. Instead, there have been attempts at oil field unitization as a management tool. Interestingly, in some groundwater basins under extreme overdraft, resource policy has also de-emphasized well-spacing requirements and moved towards basin-wide adjudication with quantity restrictions.

5. CONCLUSIONS

This chapter has presented a model for the extraction of a spatially heterogeneous, path-dependent resource by multiple spatially distributed users. The occurrence of lagged effects in such a model implies that some users may care far more about the past actions of other users than their present actions, even with discounting. In the presence of idiosyncratic effects between pairs of resource users, the optimal policy entails tradeoffs between the physical parameters of the system, individual demand functions, and the explicit spatial distribution of individual users.

Existing economic models of groundwater extraction have made assumptions about the behavior of the underlying resource that are unrealistic. In particular, the prevalence of single-cell models means that spatial aspects of policy have been entirely ignored. The assumption that groundwater is a typical common property resource drives many of the results in the existing literature. This chapter incorporates equations of motion for the state of the resource, based on the physics of water flow, into the spatially distributed groundwater extraction problem. The results shown here demonstrate that in some cases, groundwater is much closer to a private property resource than a common property resource. This is the correct physical explanation for why, at least in some cases, there may be little welfare gain from moving to an optimal extraction policy. Moreover, this analysis suggests that some of the county-level groundwater regulations observed in the real world (and ignored in previous literature) may actually be quite efficient second-best policy solutions.

Appendix A: The hydraulics of groundwater flow

Theoretical analysis of groundwater flow in the civil engineering and hydrology literature are based on the physics of water flow towards a well during pumping (for example, see Domenico, 1972, or Freeze and Cherry, 1979, for more detailed derivations of the groundwater flow equations).

Consider an extremely simple aquifer. For analytical simplicity, we assume that it has the following five properties:

1 The aquifer is horizontal.

2 The aquifer has infinite areal extent.

3 The aquifer is of constant thickness.

4 Impermeable layers above and below confine the aquifer.

5 The aquifer is homogeneous and isotropic (meaning that hydrogeological parameters are constant within the aquifer and also equal in all directions).

Before proceeding, it is necessary to define two parameters describing the physical properties of the aquifer. The storativity of a confined aquifer is the volume of water released from storage per unit of surface area per unit decrease in the hydraulic head. Storativity is dimensionless and may be thought of as the capacitance of the aquifer. The range of storativities found in confined aquifers is 0.005 to 0.00005 (Freeze and Cherry, 1979). Aquifer transmissivity is defined as the hydraulic conductivity of the aquifer multiplied by its thickness, where the hydraulic conductivity is a constant of proportionality relating specific discharge from a region to the hydraulic gradient across it. The range of values of observed transmissivities varies across over thirteen orders of magnitude from around 5×10^{-7} gal/day/ft for unfractured igneous and metamorphic rocks to around 10^8 gal/day/ft for unconsolidated gravels. Aquifers suitable for well development generally have higher transmissivities.

Theis (1935) was the first to derive an analytical solution for transient well response to pumping. In addition to the assumptions about aquifer structure described above, the Theis solution also assumes a pumping system where only a single well is pumping at a constant rate from the aquifer. Moreover, it is assumed that the well penetrates the entire depth of the aquifer, has an infinitesimal diameter, and that before the start of pumping, hydraulic head is uniform throughout the aquifer. Given a constant pumping rate Q, the drawdown of the aquifer s at any point a distance r from the well, at time t after pumping begins is defined as

$$s(t,r) = \frac{Q}{4\pi T} \int_a^\infty \frac{e^{-z}}{z} dz \tag{A. 1}$$

where

$$a = \frac{r^2 S}{4Tt} \tag{A. 2}$$

and S is the storativity, and T is the transmissivity. The integral in equation (A. 1) is the exponential integral of order one, a well-known integral whose value is given by

$$\int_x^\infty \frac{e^{-u}}{u} du = -\alpha - \ln x - \sum_{n=1}^{\infty} (-1)^n \frac{x^n}{n \cdot n!} \tag{A. 3}$$

where $\alpha = \lim_{n\to\infty} \left(1 + \frac{1}{2} + \frac{1}{3} + \cdots + \frac{1}{n} - \ln n\right) \approx 0.577216$ is Euler's constant. For notational ease, it is convenient to define the well function $w(t,r)$ where t is the time since pumping started and r is the Euclidean distance from the well, as

$$w(t,r) = \frac{1}{4\pi T} \int_{\frac{r^2 S}{4Tt}}^\infty \frac{e^{-z}}{z} dz \tag{A. 4}$$

which is an exponential integral of order one multiplied by a scaling factor. The well function is a convenient parameterization for hydrologic analysis. Note that in most hydrological literature, the well function is given without the scaling factor $1/4\pi T$. It is included within the well function here solely for simplicity of notation in the main analysis. Then, the drawdown at distance r and time t, given constant pumping rate Q, is given by $s(t, r) = Qw(t, r)$.

The Theis solution assumes a single pumping well and constant pumping rates. However, it can easily be extended to include both pumping rates that vary through time and multiple wells (for example, see Domenico, 1972). Because of linearity of the underlying transient flow equations, arithmetic summation of independent well functions can be used to calculate the drawdown through time at any point in the aquifer with multiple wells whose pumping rates vary. For example, if there are J wells pumping at constant rates $Q_1, Q_2, \ldots, Q_J$ with well j starting to pump at time t_j, then for a point that is at distances $r_1, r_2, \ldots, r_J$ from the pumping wells, drawdown at time t is given by

$$s(t, r_1, r_2, \ldots, r_J) = Q_1 w(t_1, r_1) + Q_2 w(t_2, r_2) + \cdots + Q_J w(t_J, r_J) \quad \text{(A. 5)}$$

From equations (A. 1) and (A. 5), it follows that the drawdown at any point in an aquifer depends on the location and sequence of all past pumping, so that the resource is path-dependent. The principle of superposition may also be used for the case of a single well with variable pumping rates. Assume that the initial pumping rate is Q_0, and that at times $t_1, t_2, \ldots, t_K$ this rate is incremented by $\Delta Q_1, \Delta Q_2, \ldots, \Delta Q_K$. Then the drawdown at a distance r from the pumping well at time t is given by

$$s(t, r) = Q_0 w(t, r) + \Delta Q_1 w(t - t_1, r) + \cdots + \Delta Q_K w(t - t_K, r) \quad \text{(A. 6)}$$

where the well function is zero if $t \leq t_K$. For economic analysis of groundwater extraction, equations (A. 5) and (A. 6) can be incorporated into the equations of motion for the pumping lifts in each well, given in equation (2).

Appendix B: Revisiting optimal extraction from a single-cell aquifer

Consider an aquifer in which there are J pumping agents, each with identical per-period individual benefit functions $f(u_{jt}, x_t)$. The first and second order derivatives of $f(u_{jt}, x_t)$ satisfy equivalent conditions to those in Section 3. As in Section 3, u_{jt} is the pumping of individual j during period t. However, in a single-cell aquifer, there is only one state variable, denoted here by x_t. Here, we define x_t to be the depth from the surface to groundwater, or equivalently, the pumping lift. Thus, all groundwater users, irrespective of their individual pumping, will have to pump water from the same depth. The N-period optimization problem for the aquifer is then given by

$$\max \sum_{t=1}^{N} \beta^t \sum_{j=1}^{J} f(u_{jt}, x_t) \tag{B. 1}$$

where β is the per-period discount factor, with $\beta < 1$. The equation of motion of the state variable is given by

$$x_{t+1} = x_t + \sum_{j=1}^{J} \gamma u_{jt} + R;\ t = 1, \ldots, N-1 \tag{B. 2}$$

In the single-cell aquifer, one parameter fully describes the hydrologic response of the system to pumping. This parameter is γ, and it is a constant of proportionality linking the effect of a unit withdrawal of water from the aquifer to the resultant increase in the pumping lift. Note also that per-period recharge is fixed. In the absence of pumping, there is no steady state solution to this system, and the aquifer will continue to fill towards an infinite height above the ground. This somewhat unrealistic assumption is standard within the groundwater economics literature (e.g. Brown and Deacon,1972; Gisser and Sanchez, 1980).

Equation (B. 2) may be rewritten as a summation

$$x_{t+1} = \sum_{k=1}^{t} \sum_{j=1}^{J} (\gamma u_{jk} + \widetilde{R});\ t = 1, \ldots, N-1 \tag{B. 3}$$

where $\widetilde{R} = R/J$ and, without loss of generality, we can set the initial stock level to zero. Assuming that an interior solution exists, the problem represented by equations (B. 1) and (B. 3) can be solved by the method of Lagrange multipliers. The appropriate Lagrangian is

$$L = \sum_{t=1}^{N} \beta^t \sum_{j=1}^{J} f(u_{jt}, x_t) + \sum_{t=1}^{N} \lambda_t \left\{ \sum_{k=1}^{t} \sum_{j=1}^{J} (\gamma u_{jk} + \widetilde{R}) - x_{t+1} \right\} \tag{B. 4}$$

Changing the order of the second summation in (B. 4) gives

$$L = \sum_{t=1}^{N} \beta^t \sum_{j=1}^{J} f(u_{jt}, x_t) + \sum_{t=1}^{N} \left\{ \sum_{k=t}^{N} \lambda_t \sum_{j=1}^{J} (\gamma u_{jt} + \widetilde{R}) - x_{t+1}\lambda_t \right\} \tag{B. 5}$$

From this, the first order conditions for a maximum are given by

$$\frac{\partial L}{\partial x_s} = \beta^s \sum_{j=1}^{J} \frac{\partial f(u_{js}, x_s)}{\partial x_s} - \lambda_{s-1} = 0 \tag{B. 6}$$

$$\frac{\partial L}{\partial u_{ls}} = \beta^s \frac{\partial f(u_{ls}, x_s)}{\partial u_{ls}} + \gamma \sum_{k=s}^{N} \lambda_k = 0 \tag{B. 7}$$

As before, the adjoint variable λ_t is the marginal present value shadow price of the state variable at time t. Defining the transformation $\lambda_k = \beta^k \mu_k$ where μ_k is the marginal current value shadow price at time k allows us to restate conditions (B. 6) and (B. 7) in the current value form:

$$\sum_{j=1}^{J} \frac{\partial f(u_{js}, x_s)}{\partial x_s} - \beta^{-1} \mu_{s-1} = 0 \tag{B. 8}$$

$$\frac{\partial f(u_{ls}, x_s)}{\partial u_{ls}} + \gamma \sum_{k=s}^{N} \beta^{k-s} \mu_k = 0 \tag{B. 9}$$

Noting that condition (B. 9) implies that $\partial f(u_{is}, x_s)/\partial u_{is} = \partial f(u_{js}, x_s)/\partial u_{js}$ for all i and j, we know that $u_{is} = u_{js}$. Because there is only one state variable, this means that condition (B. 8) may be rewritten as

$$J \frac{\partial f(u_s, x_s)}{\partial x_s} - \beta^{-1} \mu_{s-1} = 0 \tag{B. 10}$$

Conditions (B. 9) and (B. 10) are identical to the necessary conditions found in most simple renewable resource problems. Condition (B. 9) states that the marginal benefit of an additional unit of pumping for each groundwater user should be set equal to the shadow price of an additional unit of water held in the aquifer, multiplied by the constant of proportionality γ. In condition (B. 10), the shadow price of an additional unit of water is set equal to the aggregate marginal benefit of having one further unit of pumping lift. In this case, both of these terms will be negative.

By setting the number of time periods to infinity, the steady state condition may be obtained. At a steady state, condition (B. 10) becomes

$$\mu^* = \beta J \frac{\partial f(u^*, x^*)}{\partial x^*} \tag{B. 11}$$

Substituting this into the steady-state version of condition (B. 9) gives

$$\frac{\partial f(u^*, x^*)}{\partial u^*} + \gamma J \sum_{k=s}^{\infty} \beta^{k-s+1} \frac{\partial f(u^*, x^*)}{\partial x^*} = 0 \tag{B. 12}$$

This expression may be simplified by noting that in the steady state, $\partial f(u^*, x^*)/\partial x^*$ does not depend on time, so that the steady state condition may be simplified to

$$\frac{\partial f(u^*, x^*)}{\partial u^*} = -\gamma J \frac{\partial f(u^*, x^*)}{\partial x^*} \sum_{k=1}^{\infty} \beta^k = \frac{-\gamma J}{\delta} \frac{\partial f(u^*, x^*)}{\partial x^*} \tag{B. 13}$$

where δ is the per-period interest rate. Equation (B. 13), of course, is identical to the steady state condition obtained by using a difference equation as the equation of motion for the control variable.

Appendix C: Recovering optimality conditions for a single-cell aquifer from the general model

Although necessary conditions (7) and (9) were derived in the context of the path-dependent, spatially heterogeneous groundwater application, they are also general conditions for optimality of a wider range of resource models. For example, they can be used to recover the first order conditions for the traditional discrete time single-cell aquifer problem (e.g Brown and Deacon, 1972). This involves adding the two assumptions that define the single-cell model. First, the assumption that there is no spatial interdependency between distributed users allows the number of state variables to be reduced to one and condition (9) to be restated as

$$\frac{\partial f(u_{ls}, x_s)}{\partial u_{ls}} + \sum_{t=s}^{N} \beta^{t-s} \widetilde{\theta}(t-s) \mu_t = 0 \tag{C. 1}$$

where the new weighting function $\widetilde{\theta}(t-s)$ no longer contains spatial arguments. As a result of the reduction in the number of state variables, in condition (C. 1) the adjoint variable is analogous to the sum of all J adjoint variables in condition (9). The additional assumption of path-independence of the resource stock allows the weighting function to be passed through the summation, giving

$$\frac{\partial f(u_{ls}, x_s)}{\partial u_{ls}} + \gamma \sum_{t=s}^{N} \beta^{t-s} \mu_t = 0 \tag{C. 2}$$

The weighting function becomes a parameter, defined here as γ. It is a constant of proportionality that links a unit extraction of groundwater to the resultant change in the stock variable. Condition (C. 2) exactly reproduces necessary condition (B. 9) from Appendix B, where for reference, the single-cell resource extraction problem is solved in its entirety. With appropriate weighting functions, two-cell or multi-cell groundwater extraction problems can be recovered in a similar fashion.

Notes

1. The literature on the economics of groundwater extraction stretches back to the late 1950s and early 1960s (Milliman, 1956; Renshaw, 1963). Economic studies of groundwater extraction have followed several broad themes. Early contributions derived optimization rules for the management of groundwater resources (Brown and Deacon, 1972; Burt, 1964). Critiques of this body of work originally focused on the magnitude of the welfares difference between optimal control rules and competitive outcomes (Gisser, 1983; Gisser and Sanchez, 1980). In recent years, a further body of literature considering groundwater extraction as a differential game has emerged (Negri, 1989; Provencher and Burt, 1993).

2. Some 'two-cell' aquifer models actually contain more than two linked cells. The critical distinction between two-cell and true multi-cell models is that in two-cell models, all cells are directly connected with all others. Hence, any perturbation in one cell is immediately transmitted to all others.

3. We assume that both the number of resource users and their locations are exogenous. Incorporating endogenous well locations is beyond the scope of the current work, but for a genetic algorithm approach to a very simple well location problem, see Hsiao and Chang (2002).

4. A few existing papers have presented continuous-time models with lagged effects that are analogous to the discrete time model presented here (Kamien and Muller, 1976; Muller and Peles, 1990). These results have been used to consider such issues as optimal advertising policy and the optimal durability of products.

5. Note that the transformation to current values means that the adjoint variable equals the difference between the current value shadow price in period k and the discounted current value shadow price in period $k+1$. Hence, although μ_{jk} represents a difference in shadow prices between two successive periods, it is nonzero at a steady state.

6. Conversely, in the single-cell aquifer model, the only aggregate steady state pumping rate is that which exactly matches the per-period recharge.

References

Allen, R.C., M. and Gisser, 1984. 'Competition versus optimal control in groundwater pumping when demand is nonlinear', *Water Resources Research*, 20, 752-756.

Bredehoeft, J.D., and R.A. Young, 1970. 'The temporal allocation of groundwater - a simulation approach', *Water Resources Research*, 6, 3-21.

Brown, G., Jr., and R. Deacon, 1972. 'Economic optimization of a single-cell aquifer', *Water Resources Research*, 8, 557-564.

Burt, O.R., 1964. 'The economics of conjunctive use of ground and surface water', *Hilgardia*, 32, 31-111.

Burt, O.R., 1970. 'Groundwater storage control under institutional restrictions', *Water Resources Research*, 6, 1540-1548.

Chakravorty, U., and C. Umetsu, 2003. 'Basinwide water management: a spatial model', *Journal of Environmental Economics and Management*, 45, 1-23.

Domenico, P.A., 1972. *Concepts and Models in Groundwater Hydrology*, McGraw-Hill, New York, 405 p.

Eswaran, M., and T. Lewis, 1984. 'Appropriability and the extraction of a common property resource', *Economica*, 51, 393-400.

Feinerman, E., 1988. 'Groundwater management: Efficiency and equity considerations', *Agricultural Economics*, 2, 1-18.

Freeze, R.A., and J.A. Cherry, 1979. *Groundwater*, Prentice-Hall, Englewood Cliffs, New Jersey, 604 p.

Gisser, M., 1983. 'Groundwater: focusing on the real issue', *Journal of Political Economy*, 91, 1001-1027.

Gisser, M., and D.A. Sanchez, 1980. 'Competition versus optimal control in groundwater pumping', *Water Resources Research*, 16, 638-642.

Hsiao, C.-T., and L.-C. Chang, 2002. 'Dynamic optimal groundwater management with inclusion of fixed costs', *Journal of Water Resources Planning and Management*, 128, 57-65.

Kamien, M.I., and E. Muller, 1976. 'Optimal control with integral state equations', *Review of Economic Studies*, 43, 469-473.

Khalatbari, F., 1977. 'Market imperfections and the optimum rate of depletion of natural resources', *Economica*, 44, 409-414.

Koundouri, P., 2004. 'Potential for groundwater management: Gisser-Sanchez effect reconsidered', *Water Resources Research*, 40, WO6S16, doi:10.1029/2003WR002164.

Libecap, G.D., and S.N. Wiggins, 1985. 'The influence of private contractual failure on regulation: the case of oil field unitization', *Journal of Political Economy*, 93, 690-714.

Milliman, J.W., 1956. 'Commonality, the price system, and use of water supplies', *Southern Economic Journal*, 22, 426-437.

Muller, E., and Y.C. Peles, 1990. 'Optimal dynamic durability', *Journal of Economic Dynamics and Control*, 14, 709-719.

Negri, D.H., 1989. 'The common property aquifer as a differential game', *Water Resources Research*, 25, 9-15.

Noel, J.E., B.D. Gardner and C.V. Moore, 1980. 'Optimal regional conjunctive water management', *American Journal of Agricultural Economics*, 62, 489-498.

Noel, J.E., and R.E. Howitt, 1982. 'Conjunctive multibasin management: an optimal control approach', *Water Resources Research*, 18, 753-763.

Provencher, B., and O. Burt, 1993. 'The externalities associated with the common property exploitation of groundwater', *Journal of Environmental Economics and Management*, 24, 139-158.

Renshaw, E.F., 1963. 'The management of ground water reservoirs', *Journal of Farm Economics*, 45, 285-295.

Theis, C.V., 1935. 'The relation between the lowering of the piezometric surface and the rate and duration of discharge of a well using groundwater storage', *Transactions American Geophysical Union*, 2, 519-524.

Wiggins, S.N., and G.D. Libecap, 1984. 'Oil field unitization: contractual failure in the presence of imperfect information', *American Economic Review*, 75, 368-385.

Zeitouni, N., and A. Dinar, 1997. 'Mitigating negative water quality and quality externalities by joint management of adjacent aquifers', *Environmental and Resource Economics*, 9, 1-20.

Chapter 7

NONPOINT SOURCE POLLUTION IN A SPATIAL INTERTEMPORAL CONTEXT - A DEPOSIT REFUND APPROACH*

Renan Goetz
University of Girona, Spain

Dolors Berga
University of Girona, Spain

Àngels Xabadia
University of Zaragoza, Spain

1. INTRODUCTION

Nonpoint source pollution problems are characterized by the fact that emissions are either impossible to observe or their observation is too expensive. As a result, alternatives to the Pigouvian tax on emissions, such as an ambient tax, were proposed in order to correct these kinds of externalities (Segerson, 1988; Cabe and Harriges, 1992). However, the application of this tax in practice has been highly questioned (Horan and Shortle, 2001) as there is, among other reasons, no direct relationship between individual behavior and the amount of the ambient tax. Other approaches use the amount of applied input as a proxy of the unobservable emissions within the framework of a principal agent model (Shortle and Dunn, 1986; Dosi and Moretto, 1993; Dosi and Moretto, 1994). Yet, the amount of the purchased input is a poor approximation of the real

*The authors gratefully acknowledge the support of the Spanish Ministry of Technology research grants AGL2001-2333-C02-01 and AGL2004-00964 CICYT. They also would like to thank E. Del Rey for her helpful comments and suggestions.

emissions (Shortle and Abler, 1998). In order to relate input use and emissions more precisely, it would be necessary to have complete information about the amount of input applied at a particular location and the way the input is applied. In order to elicit this information we propose applying a deposit refund system. As seen in previous literature, this consists in a tax on the pollutant and the payment of a subsidy (refund) for the correct elimination of the pollutant. In contrast to previous literature (Sigman, 1998; Kolstad, 2000), where a pollutant was analyzed, we apply this approach to a polluting input and a contaminating byproduct. Additionally, we incorporate space to take into account the spatial heterogeneity of the land. The consideration of space, for instance in the form of a site vulnerability index, allows a more precise relationship to be established between the input use and the resulting emissions that reach the receptor, i.e. the place where environmental damage occurs.

Our proposal for the management of nonpoint source pollution is based on the figure of an authorized firm that provides the service of the correct[1] application of contaminating inputs with respect to their form and quantity. The authorized firm issues a certificate to the firm that commissioned its services. The certificate issued by the authorized firm allows the correct application of the contaminating inputs to be observed. The incorrect application, however, cannot be observed.

Given this context we must first derive the socially optimal distribution of production activities over space and their socially optimal intensity, i.e. solving the regional planner's decision problem. Due to the unobservability of the emissions that reach the final receptor or due to the presence of asymmetric information with respect to the amount and the way in which the contaminating input is applied, a policy that replicates the first-best outcome is not available. Therefore, a second-best policy based on a voluntary deposit refund system is proposed. The parameterization of space, based on site vulnerability, allows the site-specific policy to be targeted, i.e., the deposit refund system is site-specifically differentiated.

Once we have obtained the optimal distribution of production activities over space and their socially optimal intensity we have to take into account the fact that many pollutants accumulate over time. Therefore, we introduce time into our spatial model. In order to obtain the socially optimal solution over space and time we have employed a developed further two stage optimization procedure, initially introduced by Goetz and Zilberman (2000). This procedure enables us to analyze how the socially optimal spatial distribution of production activities and their corresponding optimal intensity develop over time. Given this knowledge we can design an intertemporally and spatially optimal deposit refund system.

This chapter is organized as follows: In section 2 we introduce the concept of space based on certain characteristics of the land. Based on this concept, in section 3 we present our spatial economic model and in section 4 we study the case of asymmetric information where we propose a *deposit refund system.* In section 5 the intertemporal aspects of the problem are introduced. This chapter rounds off in section 6 with some conclusions.

2. THE CONCEPT OF SPACE

We suppose that in a given region Ω, some production activities take place which cause pollution. The region Ω reflects the origin of direct emissions of a pollutant and/or the space where emissions convert into a pollutant that accumulates at a single receptor located in the region (for example, in the case of surface or underground water pollution, Ω would represent the underlying watershed). In order to have a more tractable model and to concentrate on the proposal of a deposit refund system, we will consider the case in which there is only one pollutant.[2] Since the pollutant accumulates at the receptor it will be necessary to consider time explicitly.

We first concentrated on the spatial aspect in order to analyze the repercussion of space in determining ambient policies.

We started from the idea that the region can be represented by a line that starts at point 0, the urban center of the region, and ends at point $\overline{\alpha}$, the limit of the region. Each location is identified by α, $\alpha \in [0, \overline{\alpha}]$. In order that variable α is sufficient to parameterize region Ω, we related it to variables that capture geophysical, topographical and hydrological aspects. In this way α can be interpreted as a georeferenciated index and not only as a form of coordinates. Thus, our index is a function from $\mathbb{R}^n$ to $\mathbb{R}$, where n is the amount of variables considered, and $\mathbb{R}$ is the real line. For example, α can represent a land classification system that collects the relevant characteristics of each location with respect to the pollution process at the receptor. Thus, space is not introduced in the form of a standard parameterization but in the form of a parameterization of the site vulnerability of each location within the considered region. Site vulnerability captures the extent to which the application of a contaminating input leads to an increase in the concentration of a contaminant at the receptor.

All the economic transactions in the region, except production activities take place at the urban center of the region where extension collapses at a single point since it is relatively small compared to the total region. Production activities generate emissions from a single pollutant that accumulates at a receptor located in the urban center (at $\alpha = 0$). The land outside the urban center is exclusively dedicated to production activities. The emphasis in our model is on consumers that live in the urban center, and are therefore affected by the pollutant at the receptor, rather than the ones that live outside the urban center. This is based on the hypothesis that there are many more consumers living inside the urban center than outside.

Moreover, the different productive activities are part of a competitive system where changes in regional production or changes in regional demand for inputs do not affect the production or input prices of the competitive system. In other words, they are exogenously given within our model.

To reduce the complexity of the analysis, we will concentrate on the emissions that reach the receptor (as in Hochman et al., 1977). That is, we consider the final emissions, accepting that part of the emissions from the origin are lost (absorbed, decomposed or solidified) before they reach the receptor.

Given this situation, we have determined the optimal location of the production activities in the region. With this aim, we suppose that a regional planner maximizes net actual benefits from the different agricultural activities, taking into account that an environmental standard with respect to the emissions of the pollutant that reach the receptor should not be exceeded.

However, the optimal allocation of the production activities over space is only the first stage of our optimization process over space and time. The regional planner's maximum net benefit is then reflected by a value function that depends on the optimal value of the decision variables and the exogenously defined parameters, for instance, the environmental standard with respect to the emissions of the pollutant that reach the receptor. It is precisely this parameter which becomes the decision variable in the second stage. This procedure, described in more detail by Goetz and Zilberman (2000, 2002), allows the spatial and intertemporal optimization processes to be split into two consecutive stages, this allows us to obtain an analytical solution more easily.

3. THE SPATIAL ECONOMIC MODEL

According to the separation of the spatial and intertemporal optimization procedure we start out with the first stage in which we optimize over space. For the sake of concreteness the production process is given by production activities with an infinite number of agents, say farmers. Each farmer cultivates at a given location α, where α is the parameterization of the region Ω, the total region for cultivation. Let L denote the number of hectares of arable land in Ω. Since our parameterization of space permits not only one single point, but also a subarea of Ω to correspond to a location α, the size of a given location α in relation to the size of the region Ω is captured by the density function $g(\alpha)$ with $\int_0^{\overline{\alpha}} g(\alpha)d\alpha = 1$. Thus, for each α, $g(\alpha)L$ denotes the number of hectares of arable land at location α. The support of g is the interval $[0, \overline{\alpha}]$. In the discrete case, $g(\alpha)$ denotes the proportion of Ω associated with each α.

Without losing generality, we suppose that only a single farmer cultivates at a given location α and that there is only one agricultural activity related to crop production, for instance the cultivation of wheat.[3] The share of land utilized for the cultivation of the crop at the location α is denoted by $\delta(\alpha) \in [0,1]$. The production per hectare is given by the function $f(x(\alpha),\alpha)$, where $x(\alpha)$ is the single input considered and denotes the amount per hectare of organic fertilizer applied at location α. We suppose that f presents constant returns to scale with respect to the size of the cultivated area. The production function is site-specific and therefore crop yields vary with location α. We assume that the production function f is differentiable in x and α. For simplicity of notation we denote the partial derivative of f with respect to x by f', which is usually positive, however, very high amounts of fertilizer may lead to a negative marginal productivity. Moreover, f is strictly concave with respect to fertilizer, that is, $f'' < 0$.

The fixed costs per hectare associated with the cultivation of the crop are denoted by k, which stand for the annualized capital investment costs.

The farmer at location α also has the possibility of keeping livestock. The net benefits of livestock at location α are denoted by $b(y(\alpha))$, where $y(\alpha)$ indicates the amount of manure per hectare of arable land at α, therefore being proportional to the stocking rate. The derivative of $b(y(\alpha))$ with respect to y is denoted by $b'(y(\alpha))$ with $b'(y(\alpha)) > 0$. Since we are interested in manure as a fertilizer as well as a source of water pollution and since there is a one to one relationship between the amount of manure and the number of animals, we have opted to express the number of animals in terms of manure. Farmers either use manure as organic fertilizer or if an excess exists, they dump it. Let $e(\alpha)$ denote the amount of manure in excess of organic fertilizer applied per hectare of arable land at α. The per hectare manure balance condition for the farm is given by:

$$e(\alpha) = y(\alpha) - \delta(\alpha)x(\alpha) \text{ for any } \alpha \in [0, \overline{\alpha}].$$

Moreover, the use of fertilizer is not only productive but also leads to emissions that are captured by the emission function. Organic fertilizer applied to the crop at location α leads to the emission function $\phi(x(\alpha),\alpha)$ which denotes the quantity of emissions per hectare as a result of the use of organic fertilizer at location α that reaches the receptor located at $\alpha = 0$. We assume that $\phi(x(\alpha),\alpha)$ is differentiable in x and α. We denote its partial derivative with respect to x by $\phi_x(x(\alpha),\alpha)$, with $\phi_x > 0$. The dumping of excess manure leads to the emission function per hectare $\phi^E(e(\alpha),\alpha)$, whose partial derivative with respect to e is denoted by $\phi_e^E(e(\alpha),\alpha)$ and where $\phi^E(\overline{e},\alpha) > \phi(\overline{x},\alpha)$, for all $\overline{e} = \overline{x}$. We assume that the single pollutant is given by nitrate NO_3^-.

For all α, let $p(\alpha)$ denote the output prices that are faced by the farmer located at α. We suppose that for all $\alpha > 0$, $p(\alpha)$ differs from $p(0)$ only by the transportation costs that have to be taken care of by the farmer with $p'(\alpha) < 0$. Thus, we rewrite prices as $p(\alpha) \equiv p(\varsigma(\alpha))$ where $\varsigma(\alpha)$ is a function that takes only distance into account and no other aspects of the land classification system. We also consider the application costs of the fertilizer which is defined by $c(x(\alpha))$ where $c'(x(\alpha)) > 0$ denotes its derivative with respect to x.

Let $c^E(e(\alpha))$ denote the cost of dumping of excess manure, and $c^{E\prime}(e(\alpha))$ its derivative with respect to e. We suppose that $c^E(\overline{e}) < c(\overline{x})$ for any $\overline{e} = \overline{x}$.

Given this setup and taking into account an environmental standard that introduces an upper limit on the concentration of the pollutant at the receptor, the decision problem of the regional planner, say problem (R), consists in determining the optimal value for $x(\alpha)$, selecting $\delta(\alpha)$ the optimal activity itself together with its scale, and choosing the optimal scale of husbandry $y(\alpha)$ in order to maximize the net benefits of the crop production and husbandry.

The solution to the regional planner's decision problem allows environmental policies to be designed that induce individuals to behave optimally from a regional perspective by correcting the optimal intensity and the scale of the activity. See http://www.udg.es/fcee/professors/goetz for details on the solution of problem (R) and possible environmental policies under the assumption of full information in an extended version of the chapter.

In contrast to the regulator, farmers do not take into account pollution at the receptor. To correct for this negative production externality a market intervention is required. Pigouvian taxes on emissions at the origin are considered to be insufficient to correct the externality (see Henderson, 1977; Hochman and Ofek, 1979; and Tomasi and Weise, 1994). Hochman and Ofek (1979) have shown that an adequate tax should equal the aggregate of the spatially differentiated marginal damage at each location α. Like Hochman and Ofek (1979), Goetz and Zilberman (2002) introduced a final emissions function that relates the concentration of the pollutant at the urban center with the farmers' emissions at location α.

In this way, the Pigouvian tax on the final emissions at the receptor is able to determine the optimal allocation of land use and fertilizers. However, since this tax is imposed on final emissions at the receptor it is constant through space, i.e. it is not spatially differentiated. Yet, these policies cannot be implemented because of the information required. As an alternative, input taxes (nitrogen tax) have been proposed. However, as shown by Goetz and Zilberman (2002) input taxes alone are not able to establish the social optimum. In order to achieve the socially optimal outcome the input taxes need to be complemented by land-use taxes that, depending on the curvature of the emission function, are either positive (tax) or negative (subsidy).

While land use can be observed easily by the regulator, the amount of input applied to the crop cannot be observed easily. Thus, the presence of asymmetric information impedes input taxes establishing the socially optimal outcome.

The literature has not yet developed a widely accepted solution to the problem of the optimal regulation of nonpoint source pollution.[4] All previous approaches coincide in the fact that they do not take into account the way the contaminating inputs are applied. To a great extent it is not only the amount of input which is responsible for emissions but also the way the input is applied.

In order to give farmers incentives to reveal information about the correct application of the inputs with respect to the amount and the method of its application, we propose a deposit refund system that is presented formally in the following section. This approach is new to the literature on optimal management of nonpoint source pollution. Sigman (1998) and Kolstad (2000) previously applied this approach in the literature on regulating a pollutant directly. In contrast, the analysis presented in this chapter applies this approach in order to regulate a polluting input and a contaminating byproduct with respect to the applied amount and the method of its application taking the spatial heterogeneity of the land into account. The input is applied by an authorized firm that certifies the amount and the correct form of its application at a given location. For example, certain farmers or special firms may have the approval of authorities to apply organic fertilizer. Moreover, the authorized firm issues a document that certifies the correct application of the input with respect to its form and quantity. In order to provide incentives to apply fertilizer correctly, the regulator gives a subsidy to reward the correct application. Farmers are only entitled to the subsidy if they present the certificate of the authorized firm to the regulatory body. In this way, the regulator can observe the correct application of the farm's fertilizer. The subsidy corresponds to the net savings of the social costs of correct versus incorrect application. The regulator can derive the remaining part of organic fertilizer that has not been applied correctly by observing the total amount of manure generated at location α. To reflect the social costs of the incorrect application of the organic fertilizer a tax needs to be introduced. As the net savings of the social cost for the correct application are reimbursed via the subsidy, this tax is imposed on the organic fertilizer independently of whether it has been applied correctly or incorrectly.[5]

4. SITE-SPECIFIC DEPOSIT REFUND SYSTEM

In this section we depart from the spatial economic model discussed previously. The amount per hectare of fertilizer applied correctly by an authorized firm at location α is denoted by $a(\alpha)$, while the amount per hectare of fertilizer applied by the farm itself at location α is denoted by $x(\alpha)$. Thus, $x(\alpha) + a(\alpha)$ is the total amount per hectare of fertilizer applied at location α.

At location α, the authorized firm charges $c^A\left(a(\alpha)\right)$ for the correct application of $a(\alpha)$ per hectare.[6]

Likewise, the application of $x(\alpha)$ by the farm leads to application costs of $c\left(x(\alpha)\right)$ per hectare, with $c^A\left(\overline{a}\right) > c\left(\overline{x}\right)$ for every $\overline{a} = \overline{x}$. We denote the cost function derivatives with respect to x and a by $c'(x(\alpha))$ and $c^{A\prime}(a(\alpha))$, respectively. The per hectare production function now reads as $f\left(x(\alpha) + a(\alpha), \alpha\right)$.

Like before, we denote the derivative of f with respect to either x or a by f'. The emission function can now be differentiated according to the form of the application of the fertilizer. Organic fertilizer applied by an authorized firm at location α leads to the emission function $\phi^A(a(\alpha), \alpha)$, and to $\phi(x(\alpha), \alpha)$ if the farm applies the organic fertilizer itself. We denote their partial derivatives with respect to a and x by $\phi_a^A(a(\alpha), \alpha)$ and $\phi_x(x(\alpha), \alpha)$, respectively. These emission functions satisfy $\phi_x(\cdot)$, $\phi_a^A(\cdot) \geq 0$, and $\phi(\overline{x}, \alpha) > \phi^A(\overline{a}, \alpha)$ for any $\overline{x} = \overline{a}$. In this new framework, the manure constraint per hectare is given by

$$y(\alpha) = e(\alpha) + \delta(\alpha)\left(a(\alpha) + x(\alpha)\right) \text{ for any } \alpha \in [0, \overline{\alpha}].$$

The spreading of $e(\alpha)$, the amount of excess manure dumped per hectare of arable land, leads to the emission function $\phi^E(e(\alpha), \alpha)$. Its partial derivative with respect to e is denoted by $\phi_e^E(e(\alpha), \alpha)$ with $\phi^E(\overline{e}, \alpha) > \phi(\overline{x}, \alpha)$ for all $\overline{e} = \overline{x}$ and $\phi^E(\overline{e}, \alpha) > \phi(\overline{a}, \alpha)$ for all $\overline{e} = \overline{a}$. Let $c^E(e(\alpha))$ denote the costs of spreading the excess manure and $c^{E\prime}(e(\alpha))$ its derivative with respect to e. We suppose that $c^E(\overline{e}) < c(\overline{x})$ for any $\overline{e} = \overline{x}$.

Given this setup and taking into account that the regulator has imposed an environmental standard, denoted by z, the regional planner's decision problem, referred to as (RA), is given by

$$V(z) \equiv \max_{\{a(\alpha),x(\alpha),y(\alpha),\delta(\alpha),e(\alpha)\}} \int_0^{\overline{\alpha}} \delta(\alpha)\left[p(\alpha) f\left(x(\alpha) + a(\alpha), \alpha\right) - k - \right.$$

$$\left. - c(x(\alpha)) - c^A\left(a(\alpha)\right)\right] g(\alpha) L d\alpha + \int_0^{\overline{\alpha}} \left\{b(y(\alpha)) - c^E(e(\alpha))\right\} g(\alpha) L \, d\alpha$$

subject to

$$z \geq \int_0^{\overline{\alpha}} \left\{\delta(\alpha)\left[\phi\left(x(\alpha), \alpha\right) + \phi^A\left(a(\alpha), \alpha\right)\right] + \phi^E\left(e(\alpha), \alpha\right)\right\} g(\alpha) L \, d\alpha, \tag{PC-RA}$$

$$e(\alpha) = y(\alpha) - \delta(\alpha)\left(a(\alpha) + x(\alpha)\right) \text{ for any } \alpha \in [0, \overline{\alpha}], \tag{MC-RA}$$

$$a(\alpha) \cdot g(\alpha) L \geq 0,\ x(\alpha) \cdot g(\alpha) L \geq 0,\ y(\alpha) \cdot g(\alpha) L \geq 0,\ \delta(\alpha) \cdot g(\alpha) L \geq 0,$$
$$e(\alpha) \cdot g(\alpha) L \geq 0, \text{ and } (1 - \delta(\alpha)) \cdot g(\alpha) \geq 0 \text{ for any } \alpha \in [0, \overline{\alpha}], \tag{LULC-RA}$$

where z denotes the maximal admissible concentration of the pollutant at the receptor and $V(z)$ the value function of this decision problem.

In order to analyze this problem more easily we have replaced $e(\alpha)$ by $y(\alpha) - \delta(\alpha)\,(a(\alpha) + x(\alpha))$ as defined by the manure constraint (MC-RA). In order to simplify notation we have maintained this substitution unless required for an unambiguous notation. We define the Lagrangian function $\mathcal{L}^1$ where we introduce the multipliers μ, ξ^a, ξ^x, ξ^y, ξ^δ, ξ^e and χ. The multiplier μ is associated with the pollution constraint (PC-RA), while all other multipliers are related with the lower and upper limit constraints (LULC-RA).

The argument α of the variables/functions a, x, y, δ, p, g and of the Lagrange multipliers will be omitted in order to simplify notation. The Lagrangian is therefore given by

$$\begin{aligned}
\mathcal{L}^1 = & \int_0^{\overline{\alpha}} \delta\left[pf(x+a,\alpha) - k - c(x) - c^A(a)\right] gL\, d\alpha + \\
& + \int_0^{\overline{\alpha}} \left\{b(y) - c^E(e)\right\} gL\, d\alpha + \\
& + \mu\left[z - \int_0^{\overline{\alpha}} \left\{\delta\left[\phi(x,\alpha) + \phi^A(a,\alpha)\right] + \phi^E(e,\alpha)\right\} gL\, d\alpha\right] + \\
& + \int_0^{\overline{\alpha}} \left\{\xi^a a + \xi^x x + \xi^y y + \xi^\delta \delta + \xi^e e + \chi(1-\delta)\right\} gL\, d\alpha.
\end{aligned}$$

We assume that both $g(\alpha)$ is strictly positive for all α and that a unique solution to this problem exists. Then, the solution has to comply with the following conditions at every location α:

$$\begin{aligned}
\mathcal{L}_a^1 &= \delta\left[pf'(x+a,\alpha) - c^{A\prime}(a) + c^{E\prime}(e) - \mu\phi_a^A(a,\alpha) + \mu\phi_e^E(e,\alpha) - \xi^e\right] + \\
&\quad + \xi^a = 0. \\
\mathcal{L}_x^1 &= \delta\left(pf'(x+a,\alpha) - c'(x) + c^{E\prime}(e) - \mu\phi_x(x,\alpha) + \mu\phi_e^E(e,\alpha) - \xi^e\right) + \\
&\quad + \xi^x = 0. \\
\mathcal{L}_y^1 &= b'(y) - c^{E\prime}(e) - \mu\phi_e^E(e,\alpha) + \xi^y + \xi^e = 0. \\
\mathcal{L}_\delta^1 &= pf(x+a,\alpha) - k - c(x) - c^A(a) + c^{E\prime}(e)(x+a) + \mu\phi_e^E(e,\alpha)\cdot \\
&\quad \cdot(x+a) - \mu\left(\phi(x,\alpha) + \phi^A(a,\alpha)\right) + \xi^\delta - (x+a)\,\xi^e - \chi = 0.
\end{aligned}$$

$$\mu \geq 0,\ \mu\left[z - \int_0^{\overline{\alpha}} \left\{\delta\left[\phi(x,\alpha) + \phi^A(a,\alpha)\right] + \phi^E(e,\alpha)\right\} gL d\alpha\right] = 0.$$

In order to concentrate on the economic interpretations we assume that the Kuhn-Tucker conditions related to the restriction (LULC-RA) are satisfied.[7] Once the optimal value of all decision variables are obtained, the (MC-RA) condition allows us to determine the optimal value of e.

We can now analyze the implementation of a deposit refund system to give farmers incentives to contract an authorized firm that guarantees the correct application and the correct amount of input. The following proposition defines a deposit refund system that establishes the optimal social outcome.

PROPOSITION 1 *(Site-specific deposit refund system)*
Given the existence of an authorized firm and providing that the total amount of organic fertilizer available to the farmer can be observed at each location α, an optimal policy can be obtained by the deposit refund system defined as follows:
(a) a spatially differentiated tax over the total amount of organic fertilizer applied at location α, $\tau^o(\alpha)$, equal to

$$\tau^o(\alpha) = \mu^* \phi_x(x^*, \alpha), \textit{ and}$$

(b) a spatially differentiated tax on the excess manure, $\tau(\alpha)$, given by

$$\tau(\alpha) = \mu^* \phi_e^E(e^*, \alpha), \textit{ and}$$

(c) a spatially differentiated subsidy over the amount of organic fertilizer correctly applied at location α, $s^o(\alpha)$, equal to

$$s^o(\alpha) = \tau^o(\alpha) - \mu^* \phi_a^A(a^*(\alpha), \alpha), \textit{ and}$$

(d) a spatially differentiated land-use tax or subsidy $\sigma(\alpha)$ given by

$$\sigma(\alpha) = \mu^* \left(\phi(x^*, \alpha) + \phi^A(a^*, \alpha)\right) - \tau^o(\alpha) \cdot (a^*(\alpha) + x^*(\alpha)) + \\ + s^o(\alpha) \cdot a^*(\alpha) \gtreqless 0.$$

For simplicity of notation, the argument α of the taxes and subsidies τ^o, τ, s^o, σ is suppressed unless it is required for an unambiguous notation.

Proof. These instruments are obtained straightforwardly by comparing the first order conditions (*f.o.c.*) of the regional planner problem (RA) with the *f.o.c.* of the farmers' decision problems with a deposit refund system and land-use taxes in place. The farmer's decision problem where the argument α of the variables/functions a, x, y, δ, p, g and of the Lagrange multipliers are omitted in order to simplify notation is given by:[8]

$$\max_{\{a,x,y,\delta\}} gL \left\{ \delta \left[pf(x+a, \alpha) - k - c^A(a) - c(x) - \tau^o(x+a) + s^o a - \sigma \right] + \right.$$

$$+ \left[b(y) - c^E(e) - \tau e\right] + \left[\xi^a a + \xi^x x + \xi^y y + \xi^\delta \delta + \xi^e e + \chi (1 - \delta)\right]\Big\},$$

where e is substituted by $y - \delta\,(x + a)$. Being $\mathcal{L}$ the Lagrangian of this maximization problem,

$$\begin{aligned}
\mathcal{L}_a &= \delta\left(pf'\left(a + x, \alpha\right) - c^{A\prime}(a) + c^{E\prime}(e) - \tau^o + s^o + \tau - \xi^e\right) + \xi^a = 0.\\
\mathcal{L}_x &= \delta\left(pf'\left(a + x, \alpha\right) - c'(x) + c^{E\prime}(e) - \tau^o + \tau - \xi^e\right) + \xi^x = 0.\\
\mathcal{L}_y &= b'(y) - c^{E\prime}(e) - \tau + \xi^y + \xi^e = 0.\\
\mathcal{L}_\delta &= pf\left(x + a, \alpha\right) - k - c^A(a) - c(x) - \tau^o\left(x + a\right) + s^o a - \sigma +\\
&\quad + c^{E\prime}(e)\left(x + a\right) + \xi^\delta + \tau\left(x + a\right) - \xi^e\left(x + a\right) - \chi = 0.
\end{aligned}$$

■

It is important to note that the sign of the land-use tax, σ, can be either positive or negative (Goetz and Zilberman, 2002). That is, it can be either a tax or a subsidy.

The implementation of the deposit refund system as defined in Proposition 1 requires the regulator to be able to observe the total amount of organic fertilizer for every farmer. As the regulator is informed about the amount of manure applied by the authorized firm, he knows the amount of manure applied either on the field or dumped somewhere else. However, the regulator cannot discriminate between manure that is applied by the farmer and manure in excess of the optimal intensity of production. Therefore, we propose defining a single manure tax, $\overline{\tau}(\alpha)$, that does not distinguish between $x(\alpha)$ and $e(\alpha)$. It is given by $\overline{\tau}(\alpha) = \max\{\tau(\alpha), \tau^o(\alpha)\}$ and it is applied over the part of the total amount of manure not applied by the authorized firm at location α, that is, over $\delta x + e$. At the end of the cultivating period the regulator can observe the total output. Thus, given the information about a he can estimate x. With this information, the regulator may reimburse part of the collected tax, i.e., $\left[\overline{\tau}(\alpha) - \tau^o(\alpha)\right] x$ if $\overline{\tau}(\alpha) = \tau(\alpha)$ and $\left[\overline{\tau}(\alpha) - \tau(\alpha)\right] e$ if $\overline{\tau}(\alpha) = \tau^o(\alpha)$. Consequently, the land-use tax has to be modified accordingly.

Proposition 1 establishes the conditions to determine the optimal taxes and subsidies. However, before implementing this deposit refund system the regulator has to verify that the implementation of this system is socially, and privately desirable. That is, the social and private net benefits are bigger using the deposit refund system compared to a system where no authorized firm exists. In other words, two types of inequality constraints have to be fulfilled (social implementation constraints and private participation constraints).

5. OPTIMAL LAND ALLOCATION OVER SPACE AND TIME

After having analyzed the optimal allocation of inputs and land over space and the proposal of an environmental policy to achieve a second best social outcome in the first stage, we now turn to the intertemporal optimization of the optimal spatial allocation in the second stage.

The value function $V(z)$ of the spatial allocation problem in the first stage is now employed in the second stage of the regional planner's decision problem to determine the optimal intertemporal allocation. The objective function consists in the sum of the value function $V(z(t))$ and a function $m(s(t))$ that expresses the monetary damages caused by the concentration of the pollutant at the receptor. The variable $s(t)$, the state variable in the second stage, denotes the concentration of the pollutant at the receptor. The left-hand side value of the constraint (PC-RA) of the first stage problem, z, becomes the decision variable in the second stage. It still denotes the emissions of the entire region that reach the receptor; however, it now depends on t. The terminal value function is given by $F(s(T))$. Given this context the regional planner's decision problem reads as:

$$\max_{z(t)} \int_0^T \Big(V(z(t)) - m(s(t)\Big) e^{-rt}\, dt - e^{-rT}\, F(s(T)),$$

subject to

$$\dot{s}(t) = z(t) - \xi s(t), \qquad s(0) = s_0, \qquad z(t) \in \mathcal{Z}, \tag{RT}$$

where a dot over a variable denotes the operator $\frac{d}{dt}$. The set $\mathcal{Z}$ presents the interval $[0, \bar{z}]$, where the upper limit of the set corresponds to the highest possible amount of emissions that could reach the receptor. Argument t of all the dynamic variables is dropped to simplify notations whenever possible without introducing an ambiguous notation. Hence, the current Hamiltonian value in the second stage $\mathcal{H}$ reads as $\mathcal{H} \equiv V(z) - m(s) - \psi(z - \xi s)$. Note, that a negative sign in front of the costate variable ψ has been introduced to facilitate its interpretation. The necessary conditions[9] for an interior solution $0 < z < \bar{z}$ of problem (RT) are given by

$$\mathcal{H}_z = V_z - \psi = 0 \Rightarrow \qquad \mu(t) = \psi(t) \tag{2}$$

$$\dot{\psi} = \psi r + \mathcal{H}_s = \psi(r + \xi) - m'(s(t)) \tag{3}$$

$$\dot{s} = z - \xi s\,, \qquad s(0) = s_0, \tag{4}$$

where we made use of the dynamic envelope theorem to obtain the result of equation (2). This equation states that the marginal value of the final emissions of the entire region should equal its shadow cost ψ, which, in turn, is equal to the shadow cost of the spatial allocation problem μ. Equation (3) explains the change in the shadow cost of a delayed reduction of a marginal unit of the pollution stock from period t to period $t+1$. It states that the change is equal to the extra interest and "decay" forgone paid on the shadow cost minus the cost of extra pollution associated with the delay. The transversality condition requires that

$$\psi(T) = F'(s(T)). \tag{5}$$

It shows that the shadow cost at the terminal point of time has to equal the marginal terminal value of the amount of pollutant at the receptor. Hence, the particular solution of differential equation (3) yields

$$\psi(t) = F'(s(T))e^{(r+\xi)(t-T)} + e^{(r+\xi)t}\int_t^T m'(s(\tau))e^{-(r+\xi)\tau}\,d\tau, \tag{6}$$

which states that shadow costs at time t correspond to the sum of the discounted marginal terminal value for the remaining time $(T-t)$ and the "present value" of the integral of the discounted marginal damage from time t to the end of the planning horizon. For both terms of the sum the discount rate consists of the social discount rate and the natural decay rate.

Knowing that the optimal values of $\mu(t)$ and $\psi(t)$ are identical we are able to write the dynamic version of the proposed deposit refund system, simply by replacing μ by $\mu(t)$ in Proposition 1.

6. SUMMARY AND CONCLUSIONS

For nonpoint source pollution neither the quantity nor the polluter is known. Furthermore the problem is exacerbated by the heterogeneity of the biophysical conditions that determine the transport and transformation process of the pollutant from its origin to the arrival at the receptor. As a solution to this problem this chapter proposes incorporating the dimension of space in order to target an environmental policy specific to the biophysical conditions at each location. In this way it is possible to relate the amount of contaminating input and contaminating byproduct more closely with the emissions that reach the receptor.

However, the emissions that reach the receptor do not only depend on the amount of the applied input and byproduct but also on the way these are applied. Given the fact that the regulator can neither observe the quantity nor the way the input and the byproduct are applied the problem of moral hazard exists. As a solution that is new to the literature of the management of nonpoint source pollution management, we propose a spatially differentiated deposit refund system. The results show that farmers who commission the service of an authorized firm to apply organic fertilizer correctly should receive a subsidy equivalent to the net savings of the social costs of correct versus incorrect application.

The social costs of the incorrect application of the input and the contaminating byproduct however, need to be imposed on the polluter in the form of a tax. As the net savings of the social costs for the correct application are reimbursed via the subsidy, the tax is imposed on organic fertilizer in general independent of whether it is applied correctly or not.

However, the regulator cannot observe whether the manure is used as a fertilizer by the farmer or whether it is dumped. Therefore, it is proposed to use a single tax on the part of the total manure not applied by the authorized firm. The information about the crop yields allows the regulator to reimburse part of the collected taxes after the harvest.

Apart from policy questions, the chapter also demonstrates the use of a two stage optimization technique that optimizes in the first stage the allocation of inputs and land over space and in the second stage determines how this optimal spatial allocation changes over time. In this way the optimal spatial intertemporal form of the deposit refund system can be derived.

Notes

1. Hereafter, we use the expressions "correct application" versus "incorrect application" to indicate that inputs are applied with more versus less guarantees to avoid emissions, i.e., the correct amount of input is applied and the input is applied such that emissions are minimal.

2. The results with two or more pollutants would depend heavily on the existing interactions between the different pollutants. Thus, we would obtain very specific results for each particular case.

3. Considering that more than one farmer cultivating in α would complicate the analysis without obtaining any additional insight into the problem posed. If $m > 1$ farmers cultivating in α, the functions $g^{\overline{s}}(\alpha)$ defined for each farmer $\overline{s} = 1, ..., m$ satisfies $\sum_{s=1}^{m} g^{\overline{s}}(\alpha) = g(\alpha)$.

4. For a discussion on the advantages and disadvantages of the different existing approaches see chapter by M. Khanna and R. Farnsworth of this book, and Shortle and Abler (1998).

5. An alternative to a voluntary deposit refund system, one could imagine an obligatory program where farmers would be obliged to contract the services of an authorized firm. However, this solution would not be efficient since the externality caused even by the correct application is not internalized in the farmers' decision process.

6. In order to concentrate on the proposal of a deposit refund system we assume that the net benefits of the authorized firm are zero at each period and are exogenous to the problem.

7. We use an asterisk as a superscript to denote the evaluation of a decision variable or a multiplier at its optimal value.

8. In order to simplify notation, we suppress the Lagrange multipliers ξ^a, ξ^x, ξ^y, ξ^δ, ξ^e and χ as decision variables.

9. See theorems 1 and 3 in Seierstad and Sydsæter (1987).

References

Cabe, R. and J. Herriges, 1992. 'The regulation of nonpoint-source pollution under imperfect and asymmetric information', *Journal of Environmental Economics and Management*, 22, 134-146.

Dosi, C. and M. Moretto, 1993. 'NPS Pollution, Information Asymmetry, and the Choice of Time Profile for Environmental Fees' in *Theory, Modeling and Experience in the Management of Nonpoint-Source Pollution*, edited by C.S. Russell and J.F. Shogren, Dordrecht: Kluwer Academic Publishers.

Dosi, C. and M. Moretto, 1994. 'Nonpoint Source Externalities and Polluter's Site Quality Standards Under Incomplete Information' in *Nonpoint Source Pollution Regulation: Issues and Policy Analysis,* edited by T. Tomasi and C. Dosi, Dordrecht: Kluwer Academic Publishers.

Goetz, R. and D. Zilberman, 2000. 'The dynamics of spatial pollution: The case of phosphorus runoff from agricultural land', *Journal of Economic Dynamics and Control*, 24, 143-163.

Goetz, R. and D. Zilberman, 2002. 'The economics of land zoning: the dynamic approach', *Mimeo* Universitat de Girona and University of California at Berkeley.

Henderson, J., 1977. 'Externalities in a spatial context: the case of air pollution', *Journal of Public Economics*, 7, 89-110.

Hochman, E., D. Pines and D. Zilberman, 1977. 'The effects of pollution taxation on the pattern of resource allocation: the downstream case', *Quarterly Journal of Economics*, 91(4), 625-638.

Hochman, O. and H. Ofek, 1979. 'A theory of the behaviour of municipal governments: the case of internalizing externalities', *Journal of Urban Economics*, 6, 416-431.

Horan, R.D. and J.S. Shortle, 2001. 'Environmental Instruments for Agriculture' in *Environmental policies for agricultural pollution control*, edited by J.S. Shortle and D. Abler, CABI Publishing, Oxon, U.K.

Kolstad, C.D., 2000. *Environmental Economics*, Oxford University Press, New York.

Segerson, K., 1988. 'Uncertainty and incentives for nonpoint pollution control', *Journal of Environmental Economics and Management*, 15, 88-98.

Seierstad, A. and K. Sydsæter, 1987. 'Optimal Control Theory With Economic Applications', Advanced textbooks in economics; 24; Amsterdam: North-Holland.

Shortle, J.S and D.G. Abler, 1998. 'Nonpoint Pollution' in *The International Yearbook of Environmental and Resource Economics 1997/1998,* edited by H. Folmer and T. Tietenberg, Edward Elgar, Cheltenham UK.

Shortle, J.S. and J.W. Dunn, 1986. 'The relative efficiency of agricultural source water pollution control policies', *American Journal of Agricultural Economics*, 68, 668-677.

Sigman, H., 1998. 'Midnight dumping: public policies and illegal disposal of used oil', *RAND Journal of Economics*, 29(1), 157-178.

Tomasi, T. and A. Weise, 1994. 'Water Pollution in a Spatial Model' in *Nonpoint Source Pollution Regulation: Issues and Analysis*, Chapter 7, edited by C. Dosi and T. Tomasi, Kluwer Academic Publishers, Dordrecht.

Chapter 8

TRANSBOUNDARY WATER MANAGEMENT ALONG THE U.S.-MEXICO BORDER

Linda Fernandez
University of California, Riverside, U.S.A.

1. INTRODUCTION

Forty percent of river basins worldwide are transboundary waterways shared by two or more countries (Wolf, 1999). The interdependence of the U.S. and Mexico for sharing scarce surface and subsurface water resources along a 1952 mile international border requires binational cooperation that addresses water quantity and quality issues simultaneously. The variation in waterbodies along the U.S.-Mexico border spans the variation elsewhere on earth in terms of some surface waterways flowing north to south, some flowing south to north, deep and shallow groundwater aquifers straddling the border. The U.S.-Mexico border region can serve as a laboratory for studying transboundary water processes within dissimilar societies experiencing continued expansion of shared population and commerce. Thus far, the expansion at the border is far outpacing the environmental infrastructure where approximately 12% of 16.1 million border residents lack access to safe drinking water and 57% lack access to wastewater treatment (U.S. GAO, 2000).

Economic analyses of the U.S.-Mexico border can identify incentives that lead to solutions among two or more countries sharing water across borders. Water in this border setting is a regional public good that is nonexcludable but rival in consumption. Asymmetries over time and space between the countries are important to address in such economic analyses to determine sustainable outcomes for water resources. Economic models and empirical analyses to be discussed here include investigating what leads to cooperation rather than unilateral endeavors in terms of gains and costs of arranging alternative management. Economic analyses through applied game and dynamic optimization offer guidance in how to include hydrological details to elucidate supply and demand into transboundary settings of managing water.

2. BORDER HYDROLOGY

Several watersheds offer past, present and future examples of vital binational actions to protect surface and groundwater resources. The Rio Grande, the Colorado River, and the Tijuana River are 3 of the 23 shared surface water resources along with some shared groundwater aquifers that underlie both countries. Historically, international treaties have delineated surface water quantity allocation between the U.S. and Mexico. However, excess diversions have increased while fluctuations in climate and water supply have experienced drought.

For example, the Rio Colorado Delta has been dramatically altered by transbasin diversions from the Colorado River (Morrison et al., 1996). Even though there is 1.5 million acre feet required annually for delivery to Mexico under the 1944 treaty, it is not enough. Terrestrial and marine residential and migratory flora and fauna have been altered and would need a quarter million more to stimulate natural flooding (Newcom, 2002). The remaining wetland habitat exists only where agricultural drainage water is discharged or where there is groundwater flow (Michel, 2002) and (Luecke et al., 1999). Morrison et al. (1996) argue that since Mexico receives less than 10% of Colorado River flows, it is unrealistic and inequitable for Mexico to assume all the responsibility of restoring the ecosystems in the Rio Colorado Delta to the Gulf of California. Once supported by the free-flowing Colorado River, this historical river flood plain has its water supply consumed almost entirely by upstream municipal and agricultural users. Though efforts to apply the Endangered Species Act to the species found in the wetlands habitat in Mexico have failed to date, the environmental and academic stakeholders have gained attention from other stakeholders to focus on restoration. Minute 306 (IBWC, 2000) by the International Boundary Water Commission (IBEC) is a key step for binational efforts to implement the Delta restoration slowly. The IBWC is the oldest public agency (as of 1925) responsible of binational management of water along the border.

With untreated sewage and salinity traveling across the border, there are serious water quality problems in most of the surface waterways along the border including the Tijuana River, New River, Colorado River, and Rio Grande. Urbanization in border cities next to rivers leads to urban runoff and sedimentation that is causing serious water quality problems. The Tijuana River Watershed that is binational with the upstream 2/3 of the watershed in Mexico and the remaining in the U.S. is threatened by water quality degradation. The wetlands in this watershed act as a haven for endangered and threatened bird, fish, and plant species and was recently named as a Ramsar site of international importance. Sedimentation is now gaining attention of the International Boundary Water Commission in this watershed to hold public meetings on solutions (IBWC, 2005).

The three most significant aquifers include the Hueco Bolson extending 3000 square miles in the Ciudad Juarez-El Paso region, the Mesilla Bolson extending 7450 square miles between Chihuahua and New Mexico, and the Mesa de San Luis aquifer extending 3000 square miles across Arizona and California in the U.S., and Sonora and Baja California in Mexico (Frisvold and Caswell, 2002). Existing treaties between the two countries do not regulate the distribution of groundwater between the two countries and thus any of the aquifers could be deemed an open access resource where the use of the physically shared resource is open to all (Randall, 1981).

Aquifers such as the Hueco Bolson are primarily nonrenewable if not artificially recharged. Even if one country implements conservation measures, it can still experience water shortages if the measures do not extend across the border where the aquifer is also influenced. For example, the Hueco Bolson may be experiencing some changes due to the recharge efforts by El Paso, in the U.S., but until binational coordination with Ciudad Juarez in Mexico takes place, whatever amount is recharged could be overdrawn with out proper accounting for each countries' withdrawals from the same subterranean source. The Hueco Bolson, Mesilla Bolson and other border aquifers are extracted faster than they are being recharged (Kishel, 2000). El Paso obtains about 80% of its potable water from the two Bolson aquifers as well as 20% from the Rio Grande. El Paso focuses on groundwater recharge and wastewater reclamation on the supply side. Currently, 5500 AF/year is reclaimed and the goal for 2020 is 10,000 AF/year (Hutchison, 2004). For demand side management, changing from fixed prices to increasing block rate pricing as well as implementing public education, water quantity demanded was reduced in El Paso. For example, in a normal year 40,000 AF is consumed in El Paso versus 75,000 AF/year in a drought year or 120,000 AF/year in Ciudad Juarez (Hutchison, 2004). Several measures of water quality (salinity, total dissolved solids) in the aquifers exceed levels considered safe for public health (Hayes, 1996).

So far, there is no management framework for groundwater along the border. No legal regimes or institutions currently exist for managing water quality, quantity and extraction of aquifers that cross the border. The Franco-Swiss Genevese Aquifer Agreement of 1978 between France and Switzerland regarding the Lake Geneva Basin groundwater that entails a six member commission that meets regularly to discuss usage and create annual plans for water management represents an ideal for places like the U.S. – Mexico border to strive for.

Support in the U.S. legislature during 2005 for senate bill number 1957, U.S.-Mexico Transboundary Aquifer Assessment Act, and H.R. 469 raise hopes of providing $50 million over the next decade for the first effort to map groundwater resources on the border. The period of 2005 to 2014 is specified for the mapping effort. During that time, the U.S. Geological Survey would use half of the money to characterize, chart, and assess the key aquifers in the U.S-Mexico border area. The other 50% would be distributed to border states' Water Resources Research Institutes for technological advancement. The spending level is equivalent to less than 50 cents a year for each of the more than 10 million people living along the border. The bipartisan legislation promises to reduce widespread confusion over binational water management by establishing a reliable database for decisionmaking. Mexico expresses the need to invest more in its monitoring of the Hueco Bolson and other aquifers (Mendoza, 2004).

Lack of information on border groundwater resources leads to problems of overdraft and depletion, impeding cooperation across a shared international water resource. While the consequences of dire conditions (water shortages and degraded water quality may be known) there is a lack of standardized and updated information on stock and flow conditions to enable adaptive management decisions for the dynamic changes in water. Water quantity and water quality are intertwined in most transboundary resources where drainage to water supply is hazardous if not properly treated.

The economic analyses include institutional objectives from treaties to investigate the underlying incentives to either stabilize or destabilize water among countries. Water management continues to change with involvement from federal, state, and local institutions and nongovernmental organizations on both sides of the U.S.-Mexico border. More efforts are required for the long run solutions to water quantity and quality problems. Concurrent strategies for water conservation and water reuse as well as pollution prevention in the industrial and residential sectors are mandatory for border water resources to survive.

Invasive aquatic species are causing severe damage along the U.S.-Mexico border surface waterways and have not been addressed in a water quality policy (Good Neighbor Environmental Board, 2005). Invasive species are directly consuming large volumes of scarce water, while in others their presence makes it more difficult to transport the water. In the lower Rio Grande for example, the water hyacinth and hydrilla are multiplying and choke the flow of the river (Good Neighbor Environmental Board, 2005). A floating fern, giant salvinia has become a problem on the lower part of the Colorado River. Thick mats of the plant reduce oxygen content, degrading water quality for aquatic species and curtail recreational activity such as boating and fishing and clog water intakes for irrigation.

3. HISTORICAL BINATIONAL WATER POLICY

Fischhendler et al. (2004) address linking watersheds along the U.S.-Mexico border in treaties. The linking appears in the 1944 treaty governing water quantity allocations between the two countries for the two largest surface water sources, the Colorado River and the Rio Grande. There are transaction costs in terms of time for more parties to reach consensus. Additionally, there is loss of flexibility in making adjustments in separate watersheds based on separate physical and institutional changes occurring when changes in the hydrologic cycle occur. The 1944 treaty indicated that Mexico would exchange annually 350,000 AF of Rio Grande's water for 1,500,000 AF from the Colorado River (Mumme and Aguilar, 2002). The ongoing debates regarding repayment of water quantity debts from past years leads to a sense of a necessary change from the existing agreement.

In case of extraordinary drought, the 1944 treaty provides an escape clause that enables Mexico to deliver less than the minimum in a 5 year cycle but requires it to make up the deficit over the subsequent 5 years (Article 4, paragraph B(d,1994 treaty, minute 214). Negotiations over water deficit that Mexico incurred at the end of the 1992-1997 cycle on the Rio Grande indicate that Mexico will deliver water to Texas farmers in terms of the following quantities: 14 billion gallons from Mexican tributaries and 16 billion gallons from Mexican reservoirs. The treaty is silent about drought. Even though precipitation has been below average after 1997, no formal recognition of extraordinary drought has happened to implement the escape clause in the treaty.

In 2002, the U.S. and Mexico agreed to roll the debt over to a third consecutive cycle. In 2004, Mexico repaid nearly ¾ of its remaining debt while meeting current cycle obligations (Pierson, 2004). This amount of more than 900,000 acre feet during the water delivery year ending September 30, 2004 constituted 260% of the annual average required under the treaty (Good Neighbor Environmental Board, 2005).

Although the 1944 Treaty does not explicitly mention water quality, it authorizes the International Boundary Water Commission (IBWC) to maintain adequate sanitation where waterways cross the border. The IBWC is responsible for applying boundary and water treaties between the two countries. It is in charge of dam maintenance, flood control projects, water quality monitoring and operation of international wastewater treatment plants (in Imperial Beach, CA and Yuma, AZ). These can largely be seen as engineering approaches to water problems that may require multifaceted approaches. The IBWC has reacted to border wastewater problems ex post not ex ante, partially based on the lack of explicit language regarding water quality (Mumme, 2005). The Mexican sister agency of the IBWC is the

Comision Internacional de Limites y Aguas (CILA). Minute 242 of the treaty expressly commits the U.S. to sustaining a level of salinity equivalent to the in the lowermost U.S. storage dam on the Colorado River at Imperial Dam.

The U.S.-Mexico Border Environment Cooperation Agreement better known as the La Paz Agreement from 1983, is the first of three binational efforts have provided rationale for greater financial investment at all levels (local through binational). The agreement ushered in a new era of formal binational consultation and heightened attention to environmental problems in the border region, particularly those with a clear binational component. The La Paz Agreement grandfathers the IBWC and officially acknowledges its leading role in matters related to binational water management in the border zone.

During 1995-2000, the Border 21 Program and subsequent Border 2012 Program since 2000 through present, have been developed at the national level of both countries' governments through U.S. Environmental Protection Agency (EPA) and Mexico's Secrataria de Medio Ambiente, Recursos Naturales (SEMARNAT). These programs focus on topics of air, water and land environmental quality through working groups devoted to each that identify current problems along the border. Border 2012 is focused on moving away from federally dominated organization in favor a more regional and localized series of workgroups (Mumme, 2005). Both the La Paz Agreement and the Border 21 program are executive agreements not formal treaties, so there is no formal arbitration or enforcement mechanism to implement target objectives.

The two binational environmental institutions established through the side agreement of the North American Free Trade Agreement (NAFTA), the Border Environmental Cooperation Commission (BECC) and the North American Development Bank (NADBank) are important players in border water management and the principle source of new capital investment in border water infrastructure. BECC is in charge of certifying environmental infrastructure projects for the border that address water, solid waste, wastewater, and more recently, air quality. BECC also provides technical assistance and grants for project development to border locations applying for certification on their environmental infrastructure projects. NADBank provides funding through loans and grants for the environmental infrastructure projects that BECC certifies.

4. CONTEMPORARY BINATIONAL WATER POLICY

By the end of 2004, the NADBank had disbursed $276 million in grants through the Border Environment Infrastructure Fund (BEIF), benefiting

3,810,655 people (Good Neighbor Environmental Board, 2005). At the end of 2004, with 9 years of border environmental infrastructure by the BECC and NADBank, a total of $2.1 billion worth of projects with 69 in the U.S. and 36 in Mexico for a total of 105 projects. The NAFTA environmental side agreement that established two institutions in 1994 (BECC and NADBank) provides an institutional mechanism for developing and financing border environmental infrastructure, with an emphasis on potable water supply and wastewater treatment.

BECC is structured as a truly binational agency, a single organization with representation of both countries at the board, managerial, and advisory level (Fernandez, 2004). Its technical assistance to needy communities and certification procedures for environmental infrastructure project approval has significantly improved local capabilities for accessing needed water along the border. It is integrated with federal agencies on which it most depends for financial resources, the national environmental ministries and the IBWC. There is sentiment that BECC's technical assistance and certification procedures should be modified to favor projects that directly contribute to the watershed management (Nitze, 2004). At present, in the absence of an institutional mandate to this effect, the agency officially treats each project separately, on its own merits rather than on a cumulative basis. Likewise, the IBWC and other agencies have not yet worked on an agenda for watershed management (Nitze, 2004).

Kelly and Szekely (2004) suggest joining operations of two national sections of the IBWC into a single office to improve binational coordination. The two countries' separate IBWC sections proceed differently. For example, the U.S. section has citizen advisory boards for key transboundary rivers and streams and holds public meetings regularly to interact on water issues. These boards are still developing and lack a full-scale watershed structure. It is unclear how these councils will coordinate with Mexico's river basin councils (consejos de cuencas) (Guillen, 2004). In the case of the Colorado Delta riparian conservation, a binational alliance of environmental organizations and university scholars led to the two countries agreeing on Minute 306 in August 2000 to create a binational task force to study the water requirements of the Delta ecosystem. In the westernmost portion of the border, The Tijuana River Watershed Vision Project initiated in 2003 involves diverse binational stakeholders developing a binational strategy for solving problems in the watershed. The Paso Del Norte Water Task Force (2001) started in 1999 to identify priority water issues, coordinate water and land use plans, and submit recommendations to water authorities in the U.S. and Mexico.

There has been little progress on binational groundwater management since 1973 when Minute 242, the Permanent and Definitive solution to the Salinity Crisis, endowed the IBWC with a small role in groundwater governance. The border varies in terms of surface and groundwater providing potable water sources for the burgeoning population. Tijuana relies on groundwater for potable water and its emissions to the Tijuana River affect surface water quality in San Diego County in the U.S. Increasing reliance on groundwater for municipal and industrial uses on the border has raised the potential for binational conflict over this resource.

During 2004, the NADBank allocated $7 million in grants from the water conservation fund for three canal lining projects the BECC approved to divert water solely for use in the U.S. Therefore, This could be viewed as inefficient from a binational perspective. Lining irrigation canals in the U.S. such as the All-American Canal in California prevents 0.2 million acre feet as seepage that has helped the Mesa San Luis aquifer that lowers the salinity level to Mexican agricultural activity (Frisvold and Caswell, 2002). Crop cultivation accounts for 60-80% of water use on the border (Mumme, 2005). The lining and diversions represent unilateral action by the U.S. without regard for groundwater effects on Mexico (Sanchez, 2004). The intent by California is to reduce the amount it withdraws from the Colorado River and use the water obtained from lining to sell to San Diego County. BECC approval of an increasing number of canal lining projects in the U.S. is controversial because one criterion for project approval is transboundary environmental impact. Early economic models of groundwater (Gisser and Sanchez, 1980) do not include the complexity of true problems in the border setting. For example, Gisser and Sanchez (1980) do not account for sunk costs, replacement costs, capital costs and assume recharge and extraction can be constant in a uniform private farmer optimization framework.

5. ASYMMETRY ON THE BORDER

A consistent problem along the entire U.S.-Mexico border pertains to the asymmetric financial limitations of border residents to self-financing the water management needed for the continually expanding residential population. Public finance differences take the form of San Diego's municipal budget as 27 times greater than Tijuana, the city right across the border (Frisvold and Caswell, 2002). Likewise, the wage differences between San Diego ($54 per day average) and Tijuana ($15.20 per day) are significant (Fernandez, 2005). Differences in banking, tax rules, currency conversions, budget processes and timing are problematic. Along the border of both countries, many people reside in colonias, which are unauthorized residential subdivisions in unincorporated areas of the urban fringe. They typically lack basic services of drainage, paved roads, and public utilities of electricity, water and wastewater treatment. A few colonias have been positively impacted by projects approved and financed starting in the late 1990s by public funds channeled through the NAFTA institutions BECC and

NADBank to address water supply issues (in Texas and New Mexico colonias).

Mexico faces not only physical water shortages, but also serious financial shortages for water management. Mexican municipalities need to surmount challenges of securing long term financing for vital public infrastructure projects. The Mexican tax system limits the taxation authority of local governments (Liverman et al., 1999). Under Mexican law, locally collected taxes go back to the federal government. Communities are dependent on uncertain, annual legislated appropriations for infrastructure funds (Frisvold and Caswell, 2002). This precludes border municipalities from issuing bonds or qualifying commercial loans (GAO) to finance infrastructure construction costs.

One example of a new approach to change the bond limitation involves the NADBank purchasing revenue bonds in February 2005 as a form of a loan for a solid waste project in Dona Ana County, New Mexico. The $1.5 million in bonds is in addition to nearly $100,000 in grant funds allocated to the solid waste authority to manage the project (BECC, 2004). Such financing represents an alternative to perpetual unilateral grants from the U.S. to Mexico to overcome the centralized government and financial channels in Mexico.

For example, the Border Environmental Infrastructure Fund (BEIF) is the channel for grants from the U.S. Environmental Protection Agency (U.S. Congressional Appropriations originally) that can be used in place of user fees for the first five to seven years of an environmental project (wastewater treatment plant or water purification plant) in a border municipality. Thus far, actual loans that require border municipalities to pay back finances for projects account for less than 5 percent of financing (Reed, 2000, GAO, 2000). Until now, such unilateral financing has been deemed growth inducing to perpetuate the condition that border populations grow but do not bear the costs of utilities on the border. However, a counter position is that the border economy revolves around production for consumption of goods far from the border. Thus, financing from other than the border makes sense if the consumers should be asked to bear the true costs of production. There have been attempts to widen the source of funding through programs like Promagua to foster public and private partnerships for water development (Guillen, 2004).

Mexico is centralized at the national level for water issues of any kind. Centralization is a hurdle in arranging adaptive management along the northern border. The neglect by Mexico's central government for the northern border translated into inadequate staffing and budgeting for the Mexican section of the IBWC and a slowness of response to reports on changing conditions in the border region (Kroeber, 1983). Mexican legislation has established a national system of watershed councils, initially for the larger basins such as the Rio Grande and the Colorado River to act as stakeholder interest groups. The National Water Commission (CNA) is the centralized federal agency, which must be consulted for prioritizing which projects on the border should be pursued. But, the Commission does not carry out the projects on the border. The Commission is also the entity to authorize a quantity of water allocated to local irrigation districts.

6. BORDER ECONOMIC ANALYSIS

Transboundary water resources can lead to cooperation or conflict depending on the perceptions of relative benefits to each party at the boundary (Sadoff and Grey, 2002). Economic analysis with game theory tools can highlight tradeoffs to the parties and efficiency implications of different strategies. Bennett et al (1998) indicate there is a possibility of attaining bilateral agreement on river basin management by linking river basin issues of mutual interest for benefits and costs in a repeated game.

Wagner (2001) suggests that environmental treaties (including those for water) need to include incentives for cooperation. The incentives are influenced by transfers, linkages, and/or sanctions to increase benefits over costs and make the agreement self-enforcing. Sovereignty of states precludes external enforcement of regulation by agreements, and as a result, the agreements should have the right incentives to be self-enforcing.

Cooperation in the form of joint development water projects appears in theoretical models (Barrett, 1994, and Ambec and Sprumont, 2002). The joint development approach lends itself directly to equity analysis because a cooperative agreement between the countries must be reached regarding the decision to undertake a water project and how to distribute the benefits and costs. Rogers (1997) indicates that objectives of equity and efficiency may not be met simultaneously. The potential problem with some Pareto-optimal allocations is that they might induce envy due to location of the investments (perhaps only in one country). Sadoff and Grey (2002) indicate, the redistribution of economic gains must be considered simultaneously with maximizing aggregate benefits not after, because it might be quite complex. Cooperative game theory can accommodate such strategies involving transfers among asymmetric players for fair and efficient water allocation. Barrett (1994) analyses a case of three riparian water countries through use of the Shapley value to select a unique, stable and efficient allocation rather than as a means to achieve equity. Ambec and Sprumont (2002) analyze a model of n identical riparians and allow for side payments. They find a welfare distribution that is fair and efficient.

As the number of stakeholders for analysis increases, the costs (of information and transaction) for cooperation increase. The U.S.-Mexico border has two countries, but many entities (government, private sector, civil society) involved in water issues increases costs to foster cooperation. If the water decisionmaking takes place at the watershed scale, it is possible that the various entities can cooperate. Booker and Ward (2002) examine the possibilities for binational cooperation in the upper Rio Grande basin for human water use as well as instream flows for supporting habitat requirements for the Rio Grande Silvery Minnow defined by the Endangered Species Act. The perspective is that the decision is made at the watershed level for several entities (two states in the U.S. and one in Mexico).

The model of Booker and Ward (2002) integrates hydrology, treaty rules and economic impacts in order to estimate the impacts of two policies for addressing water shortages and instream flow needs. The minnow's habitat in the upper Rio Grande basin is only 5% (100 mile stretch) of its original range down to the Gulf of Mexico (Booker and Ward, 2002).

The difference between the cooperative solution, the Pareto optimum and the private solution depends on spillover effects of benefits and costs, that are likely asymmetric. Compensation is a viable way to create incentive for cooperation and more towards symmetric benefits in this basin approach. The first of two policies examined involves reducing diversions for irrigation under low flow conditions. Lost revenues and reductions in groundwater recharge make this policy less optimal than the second. The second policy involves water market transfers in a unidirectional manner from upstream to downstream. The transfers compensate irrigators upstream for making water available for instream flows as well as future municipal use downstream. Higher instream flows result with the second policy without the severe groundwater impacts. The transfers allow for water allocated to highest value use with lower value uses compensated through sales.

When time is included in the investigation, dynamic game analysis can account for stock and flow of water quantity and quality through the state and control variables, respectively. By including a state equation as a constraint, the hydrologic details of the transboundary system can be included in the optimization decision. Early game models applied to groundwater (Negri, 1989, Provencher and Burt, 1993) discuss pumping behavior under common property arrangements and derive Markov perfect Nash equilibria where the decision rule is a function of the current value of the state variable (water stock or water pollution stock).

Game analyses can investigate whether there are gains to cooperation by comparing with noncooperation and the magnitude of the gains. It is also of interest to investigate how the gains can be redistributed in order to make cooperation sustainable. Transfers may be in other than monetary terms, such as technology as a way to foster cooperation.

Even where resource flows occur unidirectionally, precluding mutual control over negative externalities, there is the possibility of Pareto-admissable or "win-win" solutions. The upstream country imposes a negative unidirectional water burden upon the downstream country by preventing the latter from reaching an unconstrained optimal water quantity. In light of the undesirable properties of the noncooperative solution, one must identify alternative criteria that can enhance basinwide economic efficiency and satisfy reasonable notions of equity. The principle of "reasonable and equitable usage" which holds that each water user is entitled to a "fair" share of water provides a more politically feasible basis for Pareto-Admissible transboundary water sharing accords such as the Helsinki Rule (Beach et al., 2000).

Fernandez (2005) investigates optimal strategies for solving unilateral suspended sediment flow in surface water from south in Mexico to north in the U.S. within the Tijuana River watershed. Asymmetry in budgets, abatement costs, and damages between upstream and downstream are included in the empirical analysis. Incentives for controlling sediment consist of preventing property damage upstream and protecting habitat, public and environmental health downstream.

The differential game analysis of Fernandez (2005) shows cooperation is viable according to several allocation rules that include financial and technical assistance transfers from downstream to upstream through viable channels of the BECC and NADBank discussed in Section 3. Through cooperation, the watershed is optimized at a single, joint level of decisionmaking for both countries. The Shapley value, the Chander Tulkens rule, and the Helsinki Rule of reasonable and equitable sharing are among the allocation rules in the study. The Shapley value allocates as a function of the average marginal contribution by each country to net gain from cooperation. The Chander Tulkens cost sharing rule is similar to a Kaldor Hicks criterion. The rule implies the proportion of savings in costs from cooperation that a country receives is equal or greater to what the country could achieve under noncooperation (Chander and Tulkens, 1992). The Helsinki Rule recognizes the riparian countries' sovereign right within its territory but restricts it use of transboundary water to ensure reasonable and equitable shares for other riparians (Rogers, 1997). The Helsinki Rule is based on allocating according to characteristics of population and size of project among other information. The amount of finances and technical assistance devoted to Baja California for helping control upstream sedimentation is lower than it should be in all cases and the avenue to allocate the transfers is through the NAFTA institutions. Sediment stock is at its lowest level with cooperation through transfers by the Chander Tulkens and Shapley value.

Frisvold and Caswell (2000) pose a static bargaining cooperative game with the IBWC as the decisionmaker with both the U.S. and Mexican sections negotiating. A Nash solution is defined using hypothetical payoff and disagreement terms for achieving a drinking water quality standard at least cost. There are examples Frisvold and Caswell (2000) discuss that differ in manner of how costs and benefits are allocated for wastewater treatment plants. For example, for the 1951 wastewater plant in Ambos Nogales between Arizona in the U.S. and Sonora in Mexico, the IBWC recommended allocating costs in proportion to benefits where the downstream U.S. had higher benefits and therefore bore higher costs (Frisvold and Caswell, 2000).

In 1984 the Reagan administration imposed the equal finance rule for payment. Frisvold and Caswell (2000) assert that the benefit/cost ratio would have to be much higher in order for any large projects to be built under the equal finance rule. Otherwise, noncooperation would tend to be the outcome. Frisvold and Caswell (2000) use the example of sewage treatment problems in Tijuana to show that the equal cost rule impeded a cooperative solution. Mexico did not favor paying $730 million for its equal share of costs to treat the sewage that flows in the Tijuana River north into San Diego County. Instead, Mexico's noncooperative solution was to build a smaller, less expensive plant, too small to handle the total sewage for the transboundary area.

The equal cost sharing rule was dropped in 1990 and affected an international treatment plant for the Tijuana and San Diego region (Minute 283, IBWC, 1990). Mexico's cost under Minute 283 was assigned to equal the amount Mexico would have spent on its own second treatment plant (IBWC, 1990). The cost allocation is parallel to a rule applicable to international agreements (Chander/Tulkens). The rule means the cost for Mexico to cooperate (in this case for the upstream source of water pollution) is the same as noncooperation. Thus, cooperation is possible and results in higher joint benefits to both upstream and downstream countries.

Similarly, along the Rio Grande that traverses between Laredo in the U.S. and Nuevo Laredo in Mexico, the sewage flow raises the need for wastewater treatment. The U.S. is paying Mexico for the incremental cost to expand the Nuevo Laredo wastewater treatment plant to meet U.S. water quality standards rather than to act in a noncooperative way and only treat wastewater in Laredo (Frisvold and Caswell, 2000). The Nuevo Laredo arrangement is viewed as the most cost effective alternative because it truly addresses a binational water pollution problem in a binational way, not unilateral way (Frisvold and Caswell, 2002). These cities are affected in a symmetric way with the Rio Grande flow versus the Tijuana River south to north flow. The incremental costs are at 25% and do not exceed the ceiling of 33% of total actual cost according to Minute 297 (IBWC, 1997). Industries in both cities face pretreatment requirements to alleviate some financial burden for only treating at the wastewater treatment plant. Minute 297 allows for adaptive management as both countries assess operating costs and water quality standard differences between both countries after 2005.

The U.S. payments may decline if standards between the countries move closer towards the same levels.

A differential game model that links transboundary water pollution to trade policy is developed by Fernandez (2002) in the context of solving wastewater pollution problems in the Rio Grande between the border cities of El Paso in the U.S. and Ciudad Juarez in Mexico. The analysis includes a comparison of noncooperative and cooperative wastewater treatment decisions with and without NAFTA that represents trade liberalization between the two countries sharing the Rio Grande border waterway. Since the BECC and NADBank were set up through the NAFTA side agreement, they offer a tangible institutional framework for both countries to cooperate on wastewater treatment. With trade liberalization through NAFTA, the quota on cotton production is removed and offers an incentive for Mexico to reclaim wastewater. The U.S. also reclaims and reduces the wastewater pollution in the Rio Grande. The empirical results suggest cooperation and trade liberalization improve water quality in the shared border waterway.

Fernandez (2004) provides the only econometric analysis of revealed preferences of the BECC, representing a trade and environment institution through the NAFTA. The study provides quantitative tests for whether the BECC operates according to the polluter pays principle, equity, or cost minimization over eight years of decisionmaking on environmental infrastructure projects along the border. Results show that BECC favors water and wastewater projects in upstream locations that are truly alleviating transboundary pollution. In addition, the study finds the locations of the projects do not bear the majority of the project costs. Instead, binational grants finance most BECC projects. The U.S. has received 68% of the projects and virtually the same share of the grants, whereas Mexico share has been 32%. Evidence of asymmetry in water flow and credit access between north (U.S.) and south (Mexico), confirm that equity is not a goal of the BECC in the asymmetric allocation of projects on the border.

7. BEYOND THE U.S.-MEXICO BORDER

Asymmetry between countries can raise special issues to accomplish equity. Issue linkage may reverse the position of the upstream and downstream countries thereby producing a more flexible stance for binational strategy (Sadoff and Grey, 2002). If trade is an issue water negotiations are linked to, the possible strategies are involved. In an aggregated analysis with data beyond the U.S.-Mexico border from the United Nations' Global Environmental Monitoring System (GEMS) on water quality in international rivers, Sigman (2002, 2004) finds evidence of lower water pollution in rivers shared between countries with more extensive trade. But, using pollution levels rather than explicit agreements means the trade may affect the uncoordinated equilibrium as well as the results of policy coordination. Cooperation is harder to test in the preceding rivers empirical case. Linking trade and water pollution in rivers could mean that production of traded goods occurs near the border with a trading partner, as between the U.S and Mexico, then rivers that cross the border may have higher pollution with more trade. This scale effect will yield a positive link between exports and pollution, tending to counteract the policy coordination effects. As a result, any negative association between trade and pollution in the empirical analysis may understate the true effects of trade on coordination. The linkage may formally establish channels for financial and other resource transfers or side payments to take place as Kaldor Hicks compensation for a Pareto optimal outcome. The payments can be monetary or in the form of trade preferences.

While the benefit of each country depends on domestic emissions only, the cost in terms of pollution depends on domestic and upstream emissions. The literature does not include enough analysis with stochastic models nor asymmetric information between countries sharing water. Maler (1990) models theoretically two country downstream pollution in which the upstream country has private information on abatement and the downstream country has private information on damage costs. Maler shows that the outcome of bargaining among the two countries on pollution control and side payments is efficient relative to the true cost and benefit functions. Crucial for this result is Maler's assumption that the two countries have equal bargaining power in their negotiations on emission reduction. Asymmetric information about extractions known only to users under unmetered conditions complicates efficient water allocation. The price of water as an input in this case may be derived if the output can be observed. Tsur and Zemel (1995) suggest it is a combination of implementation costs and asymmetric information that requires use of mechanism design theory to define efficient groundwater allocation and prices.

Rogers (1997) examines the Ganges-Brahmaputra basin as an example of an international river basin shared between India, Bangladesh and Nepal where a cooperative solution improves all countries' welfare over a noncooperative solution. The paper investigates supplementing the binational Ganges River Treaty of 1996 between India and Bangladesh through a second water arrangement with Nepal. The Ganges Treaty provides a formula for diversion rates as a function of the flow rate during dry months (Rogers and Harshadeep, 1997). Roger's analysis involves joint optimization and compares allocations based on meeting feasibility, Pareto Admissibility and individual rationality using game concepts and engineering data of the water system. The Chander Tulkens, Shapley value and reasonableness criteria are assessed separately for the cooperative solution. In all cases, India garners the most water with the other two countries (Nepal and Bangladesh) dividing the remainder. The trilateral relations in this watershed could be modeled by a Stackelberg leader-follower game to determine the optimal share of water diverted by India, both with and without the second water arrangement. India would be the leader and Bangladesh the follower where water transfer from a third country could be analyzed. The Bangladesh may be worse off whereas India is better off.

Dufournaud (1982) investigates the Lower Mekong Basin with three riparians that the cooperative scheme was never taken seriously even though the benefits to cooperate exceeds the noncooperative alternatives. In the analysis the game approach does not depend on rational behavior and Dufournaud (1982) identifies when one riparian will abandon the agreement because it is advantageous at that point. The spatial and temporal differences leaves uncertainty surrounding these benefits to the point that cooperation would be questionable. Working out the Shapley value for subsidizing costs to cooperate could be the compensation to drive cooperation.

8. DATA NEEDS

A formal agreement on water resources data compilation and sharing is needed at the U.S.-Mexico border and most likely elsewhere. Information and accessibility gaps on water might be overcome if such an agreement exists. At the federal level under IBWC minute 289, sections of the IBWC in both countries implemented a water quality study of the Rio Grande that involved the U.S. EPA, Texas Natural Resources Conservation Commission, and both the National Water Commission and Secretaria de Desarrollo Social in Mexico (IBWC, 1994). The study was conducted in three phases beginning in 1992 and raised fundamental questions regarding protocols for standardized methods of data collection and units of measure in both countries' shared waterways. Such standardization is mandatory in order to be able to have comparable data across space and time. The second and third phases focused on regions of concern identified in the first phase. Under protocols similar to those for the Rio Grande, the IBWC coordinated a study for the Lower Colorado and New Rivers (Good Neighbor Environmental Board, 2005).

It is often the case that the water quantity maps depict measures for each country or border state only, implying transboundary aquifers stop at the border (Mendoza, 2004). This would be a limitation on baseline data for coordinated flow and stock measures to sustain the resources, because there will not be the cumulative accounting of extraction and water use rates from both countries affecting the remaining stock.

One data collection study addressing groundwater from a binational perspective began in 1997 for El Paso-Ciudad Juarez that share two major aquifers. Information on the basic data needs was provided that initial year and a second binational project was completed in 2002, Simulation Groundwater Flow in the Hueco Bolson, and Alluvial-Basin Aquifer System near El Paso, Texas that consisted of two compatible models for the U.S. and Mexico. The third project in this series was also carried out in 2002 involving a hydrogeologic and water quality study of the Hueco Bolson aquifer.

In November 2004, the Transboundary Aquifers of the Americas Workshop took place in El Paso at the International Boundary Water Commission. Recommendations that resulted included improved sharing of water supply, demand and quality information in the form of data, models and forecasts. Most of the transboundary aquifers, two different groundwater systems govern management. For example, in the Mimbres Basin of New Mexico (U.S.) and Chihuahua (Mexico), the state engineer for New Mexico is in charge and the National Water Commission (CAN) for Mexico develops plans for their own country's use. Such efforts are also receiving attention at the international level with recent contributions to enhance the Global Environmental Monitoring System/Water (GEMS/Water) Program of the United Nation Environment Program (UN News Center, 2005). The International Atomic Energy Agency (IAEA) is the latest to help contribute to the transboundary water GEMS data gathering effort.

Water resources are stochastic in supply and demand. Temporal and spatial data can indicate a probability distribution of some functional relationship of randomness is key for future modeling efforts of economists and policymakers. A transboundary setting exacerbates the stochastic nature. It would be useful to base the cost functions and benefit functions on the stochastic distribution of the states of nature (weather, market shocks).

9. CONCLUSION

Transboundary water resources require binational and multinational management for longevity and support of human and environmental uses dependent on these resources. Directions of flow in surface water and spatial dimensions of aquifers straddling international boundaries raise physical definitions from which to study incentives (economic, political, social) for coordinating water management across boundaries.

This paper has focused on the U.S.-Mexico border for addressing a variety of transboundary surface and groundwater issues. The physical setting varies in direction of surface flow and groundwater scale as much as worldwide and therefore offers points to ponder for addressing transboundary water resources anywhere. Efforts to model the hydrology, institutions, and environmental costs and benefits in economic models have been made, but are not finished in terms of solving all issues. There are gains to be made in the modeling and data gathering in order to move research and guidelines for policy forward.

Economic analyses of incentives can address asymmetries due to location, budget constraints, abatement costs and damages. The preceding discussion of various economic analyses indicate financial mechanisms are modeled based on institutions that exist for countries sharing transboundary water resources. In addition, sharing rules of the Shapley value, Chander Tulkens cost sharing rule and the Helsinki Rule on reasonableness offer guidance in determining the asymmetric allocations to make cooperation happen in the transboundary setting. On the U.S.-Mexico border there are attempts to broaden the mechanisms beyond grants to environmental bonds among other forms.

Game theory models have been developed and empirically applied as described above to transboundary surface waterways. More research can be done empirically in various settings worldwide for surface and groundwater. The unidirectional game analyses from the U.S.-Mexico border and analyses from other settings identified Section 7 indicate that the upstream can be induced to assume the bulk of costs in anticipation of gaining a significant large share of the positive economies of scale (greater security for both states' water uses). For example, if the upstream state needs to impound more water in the downstream's irrigation system in order to meet the former's hydropower production requirements, the downstream may have incentive to bear a share of water storage costs.

More economic analyses of the transboundary water issues are needed. Theoretical as well as empirical studies can offer insight to guide policy and strengthen understanding of incentives and viable strategies to manage precious water resources worldwide. The decision process is better understood and its outcome more predictable when data about the joint decisions of those who share the transboundary water resources are matched with the components of the model.

References

Ambec, S. and Y. Sprumont, 2002. 'Sharing a river', *Journal of Economic. Theory*, 107, 453-462.

Barrett, S., 1994. 'Conflict and Cooperation in Managing International Water Resources', *World Bank Policy Research Working Paper 1303*, Washington, D.C.

Beach, H, J. Hamner, J. Hewitt, E. Kaufman, A. Kurki, J. Oppenheimer, A. Wolf, 2000. *Transboundary Freshwater Dispute Resolution*, United Nations University Press, New York.

Bennett, L., S. Ragland and P. Yolles, 1998. 'Facilitating International Agreements through an Interconnected Game Approach: The Case of River Basins' in: *Conflict and Cooperation on Transboundary Water Resources*, edited by R. Just and S. Netanyahu, Kluwer Academic Publishers, Boston.

Booker, J. and F. Ward, 2002. 'Restoring Instream Flows Economically: Perspectives from an International River Basin' in *Both Sides of the Border: Transboundary Environmental Management Issues Facing Mexico and the United States*, edited by L. Fernandez and R. Carson, Kluwer Academic Publishers, Dordrecht.

Border Environmental Cooperation Commission and North American Development Bank, 2004. *Joint Status Report, June 30, 2004* BECC, Ciudad Juarez, available at http://www.cocef.org.

Chander, P. and H. Tulkens, 1992. 'Theoretical foundations of negotiations and cost sharing in transfrontier pollution problems', *European Economic Review*, 36, 388-398.

Dufournaud, C., 1982. 'On the mutually beneficial cooperative scheme: Dynamic change in the payoff matrix of international river basin schemes', *Water Resources Research*, 18, 764-772.

Fernandez, L., 2002. 'Trade's dynamic solutions to transboundary pollution', *Journal of Environmental Economics and Management*, 43, 386-411.

Fernandez, L., 2004. 'Revealed preferences of an international trade and environment institution', *Land Economics*, 80(2), 224-238.

Fernandez, L., 2005. 'Coastal watershed management across an international border in the Tijuana river watershed', *Water Resources Research*, 41, W05003, doi:10.1029/2004WR003209.

Fischhendler, I., E. Feitelson, D. Eaton, 2004. 'The short-term and long-term ramifications of linkages involving natural resources: The U.S.-Mexico transboundary water case', *Environment and Planning C: Government and Policy*, 22(5), 633-650.

Frisvold, G. and M. Caswell, 2000. 'Transboundary water management game-theoretic lessons for projects on the U.S.-Mexico border', *Agricultural Economics*, 24, 101-111.

Frisvold, G. and M. Caswell, 2002. 'Financing Bilateral Water Projects on the U.S.-Mexico Bovrder: Past, Present and Future' in *Both Sides of the Border: Transboundary Environmental Management Issues Facing Mexico and the United States*, edited by L. Fernandez and R. Carson, Kluwer Academic Publishers, Dordrecht.

Gisser, M. and D. Sanchez, 1980. 'Competition versus optimal control in groundwater pumping', *Water Resources Research*, 16, 638-642.

Good Neighbor Environmental Board, 2005, *Water Resources Management on the U.S.-Mexico Border, Eighth Report to the President and Congress of the U.S.*, February 2005. EPA-130-R-05-001.

Guillen, R., 2004. 'Acuifero de Bolson del Hueco: Aspectos Legales e Institutionales', Conference proceedings from Transboundary Aquifers of the Americas, UNESCO/OAS ISARM Conference, Nov. 2004, El Paso.

Hayes,D., 1996. 'Drinking Water Quality in the U.S.-Mexico Border Region Assessment of Water Supplies, Treatment and Distribution', Proceedings of the Technical Conference of the Southwest Center for Environmental Research and Policy, Southwest Center for Environmental Research and Policy, San Diego, CA.

Hutchison, B., 2004. 'Hueco Bolson Groundwater Conditions and Management', Conference proceedings from Transboundary Aquifers of the Americas, UNESCO/OAS ISARM Conference, Nov. 2004, El Paso.

International Boundary Water Commission (IBWC), 1990. *Conceptual Plan for the International Solution to the Border Sanitation Problem in San Diego, CA/Tijuana, Baja California,* Minute 283, El Paso, TX, July 2, 1990.

International Boundary Water Commission (IBWC), 1994. *Binational Study Regarding the Presence of Toxic Substances in the Rio Grande/Rio Bravo and its Tributaries along the Boundary Portion between the U.S. and Mexico,* Final Report, El Paso, TX, Sept. 1994.

International Boundary Water Commission (IBWC), 1997. *Operations and Maintenance Program and Distribution of Its Costs for the International Project to Improve the Quality of the Waters of the Rio Grande at*

Laredo, TX-Nuevo Laredo, Tamaulipas, Minute 297, Ciudad Juarez, Mexico, May 31, 1997.

International Boundary Water Commission (IBWC), 2000. *Conceptual Framework for Unites States-Mexico Studies for Future Recommendations Concerning Riparian and Estuarine Ecology of the Limitrophe Section of the Colorado River and its Associated Delta*, Minute 306, Dec. 12, 2000, El Paso, TX.

International Boundary Water Commission (IBWC), 2005. Public Meeting to Discuss Erosion Control and Storm Water Management Project in Tijuana River Watershed, BECCNET Press Release, February 28, 2005, Ciudad Juarez.

Kelly, M. and A. Szekely, 2004. Modernizing the International Boundary Water Commission, Policy paper #1, Center for Latin American Studies, University of California, Berkeley.

Kishel, J., 2000. 'Lining the all-American canal: Legal problems and physical solutions', *Natural Resources Journal*, 33, 153-176.

Kroeber, C., 1983. *Man, Land and Water: Mexico's Farmlands Irrigation Policies*, University of California Press, Berkeley.

Liverman, D., R. Varady and O. Sanchez, 1999. 'Environmental issues along the United States-Mexico border: Drivers of change and the response of citizens and institutions', *Annual Review of Energy and the Environment*, 24, 607-643.

Luecke, D., J. Pitt, C. Congdon, E. Glenn, C. Valdes and M. Biggs, 1999. A Delta Once More: Restoring Riparian and Wetland Habitat in the Colorado River Delta, Environmental Defense Fund, Wash.D.C.

Maler, K., 1990. 'International environmental problems', *Oxford Review of Economic Policy*, 6, 80-108.

Mendoza, L., 2004. 'Acuifero del Bolson del Hueco: Aspectos Internacionales', Conference proceedings from Transboundary Aquifers fo the Americas UNESCO/OAS ISARM Conference, Nov. 2004, El Paso.

Michel, S., 2002. 'The Geography of Water Transfers and Urbanization in Baja and Southern California' in *Both Sides of the Border:*

Transboundary Environmental Management Issues Facing Mexico and the United States, edited by L. Fernandez and R. Carson, Kluwer Academic Publishers, Dordrecht.

Morrison, J., S. Postel, and P. Gleick, 1996. *The Sustainable Use of Water in the Lower Colorado River Basin*, The Pacific Institute, Oakland.

Mumme, S., 2000. 'Minute 242 and beyond: Challenges and opportunities for managing transboundary groundwater on the Mexico-U.S. border', *Natural Resources Journal*, 40, 341-378.

Mumme, S. and I. Aguilar, 2002. 'Managing Border Water to the Year 2020: The Challenge of Sustainable Development', in S. Michel (ed.) Binational Water Management Planning, Southwest Consortium for Environmental Research and Policy, Monograph No. 5, San Diego.

Mumme, S. 2005. 'Improving Binational Water Management at the U.S.-Mexico Border: Treaty Compatible Institutional Opportunities and Options', Presentation at the Border Water Conference, Jan. 27-29, 2005, Center For U.S.-Mexico Border Studies, San Diego.

Negri, D., 1989. 'The common property aquifer as a differential game', *Water Resources Research*, 25, 9-15.

Newcom, J., 2002. The Colorado river: Coming to consensus, *Western Water*, March/April 2002, Water Education Foundation, Sacramento.

Nitze, W., 2004. Draft: The Role of Climate Change in Water Management in the U.S.-Mexico Border Region: A Challenge for the BECC, the NADBank and Other Binational Institutions, Gemstar Group, Wash. D.C.

Paso del Norte Water Task Force, 2001. 'Water Planning in the Paso del Norte: Toward Regional Coordination', report to Border 21 Water Working Group, Ciudad Juarez, available at: http://www.sharedwater.org/documents/water_task_force_report.pdf

Pierson, E., 2004. 'Mexico Chips Away at Water Debt While U.S. Tries for Another Payment', The Brownsville Herald, July 11, 2004, http://www.rioweb.org/archive/jss8-wbr071104.html.

Provencher, B. and O. Burt, 1993. 'The externalities associated with the common property exploitation of groundwater', *Journal of Environmental Economics and Management*, 24(2), 139-158.

Randall, A., 1981. 'Property entitlements and pricing policies for a maturing water economy', *Australian Journal of Agricultural Economics*, 25, 195-212.

Reed, C., 2000. 'Expanding the Mandate: Should the Border Environmental Cooperation Commission and North American Development Bank Go Beyond Water, Wastewater and Solid Waste Management Projects and How do They Get There?', Texas Center for Policy Studies, Austin.

Rogers, P., 1997. 'International River Basins: Pervasive Unidirectional Externalities' in *The Economics of Transnational Commons*, edited by P. Dasgupta, P., Maler, and A. Vercelli, Clarendon Press, Oxford.

Rogers, P. and N. Harshadeep, 1997. *The Farakka Ganges Water Sharing Treaty: Some Implications for India and Bangladesh*, ASTEN, The World Bank, Wash.D.C.

Sadoff, C. and Grey, D., 2002. 'Beyond the river: The benefits of cooperation on international rivers', *Water Policy*, 4(5), 389-403.

Sanchez, V., 2004. *Revestimiento del Canal Todo Americano, Colegio de la Frontera*, Tijuana.

Sigman, H., 2004. 'Does trade promote environmental coordination? Pollution in international rivers', *Berkeley Electronic Journal in Economic Analysis and Policy*, 4 (2), September.

Sigman, H., 2002. 'International spillovers and water quality in rivers: Do countries free ride?', *American Economic Review*, 92, 1152-1159.

Treaty Regarding Utilization of the Waters of Colorado and Tijuana Rivers and of the Rio Grande, signed Feb. 3, 1944, U.S. Mex, 59 Stat. 1219, *Treaty Series 994*, U.S. Government Printing Office, Wash. D.C., available at:
http://www.usbr.gov/lc/region/pao/pdfiles/mextrety.pdf

Tsur, Y. and A. Zemel, 1995. 'Uncertainty and irreversibility in groundwater resource management', *Journal of Environmental Economics and Management,* 29(2), 149-61.

UN News Centre, 2005. 'UN Nuclear Energy Agency Helps Countries Safeguard Shared Water Sources', May 24, 2004, New York.

U.S. General Accounting Office (GAO), 2000. U.S.-Mexico Border: Despite Some Progress, Environmental Infrastructure Challenges Remain, GAO/NSIAD-00-26.

Wagner, U., 2001. 'The design of stable international environmental agreements: Economic theory and political economy', *Journal of Economic Surveys*, 15(3), 377-411.

Wolf, A., 2004. International Freshwater Treaties Database, http://www.transboundarywaters.orst.edu

Chapter 9

IRRIGATION, WATER QUALITY AND WATER RIGHTS IN THE MURRAY DARLING BASIN, AUSTRALIA

Stephen Beare
Australian Bureau of Agricultural and Resource Economics, Australia

Anna Heaney
Australian Bureau of Agricultural and Resource Economics, Australia

1. INTRODUCTION

A central tenet of water policy reform in Australia is to establish property rights to facilitate trade, allowing water to be directed to its highest valued use. The potential economic benefits from water trade are significant. In the Murray Darling Basin of Australia, which accounts for more than 70 per cent of Australian irrigated agriculture, it is estimated that interregional and interstate trade in water could increase agricultural returns by 5 per cent or almost $AU50 million per annum (Hall et al., 1994; MDBC, 2001). Perhaps more importantly, an effective water market is essential to meet increasing demand for water to fulfill environmental management objectives at least cost. It was estimated that water trade could reduce the costs of sourcing additional environmental flows from consumptive uses by nearly one third (Heaney et al., 2002). However, the physical and economic environment in which water property rights are being introduced is complex.

Defining water property rights that account for the full economic costs and benefits of water use has proven difficult. Irrigation practices can have an impact on the volume and quality of water available to downstream users and the riverine environment more generally. The extent to which trade will lead to efficiency gains depends, in part, on how well these rights account for the externalities associated with irrigation.

The main water quality issue in the Murray River system has been increasing river salinity. The results of a salinity audit, released by the Murray Darling Basin Ministerial Council in 1999, suggest that salt mobilisation in the basin would double from 5 million tonnes a year in 1998 to 10 million tonnes in 2100. The audit also reported that the average salinity of the Murray River at Morgan, upstream of the major off-takes of water to Adelaide, South Australia, will exceed the 800 EC[1] World Health Organisation threshold for desirable drinking water quality in the next 50 to 100 years (MDBMC, 1999). As a substantial proportion of the salt load in the Murray River is due to return groundwater flows from irrigation, changes in irrigation water use through trade or improvements in irrigation efficiency have the potential to contribute to mitigating the problem of increasing river salinity.

To evaluate salinity management options in the Murray Darling Basin more generally, the Salinity and Landuse Simulation Analysis (SALSA) model was developed as a simulation modeling framework that incorporates the interrelationships between land use, vegetation cover, surface and groundwater hydrology and agricultural returns. This model was developed at ABARE, in cooperation with the Commonwealth Scientific and Industrial Research Organisation (CSIRO). Initially the model was used to examine the impact of targeted land use changes to reduce saline groundwater discharge from dryland agriculture (Heaney et al., 2000). The model was developed further to incorporate irrigation management options in the Riverland region of South Australia allowing estimates of the benefits of improved irrigation efficiency as a tool for salinity management in irrigation areas (Heaney et al., 2001). Here, the model is extended to examine salinity mitigation options in the irrigation areas of the Murray River and its major southern tributaries. The geographic area under consideration is shown in figure 1. The irrigation areas under consideration are listed in Table 1, in upstream to downstream order.

The outline of the remainder of the chapter is first to provide some background on irrigation, salinity and water policy within the Murray Darling Basin. This is followed by a description of the model and methods used for calibration. The design of the reference case simulation and two alternative scenarios used to explore the impacts of water reallocation, trade and improvements in water use efficiency are then presented. Concluding comments follow a discussion of the results.

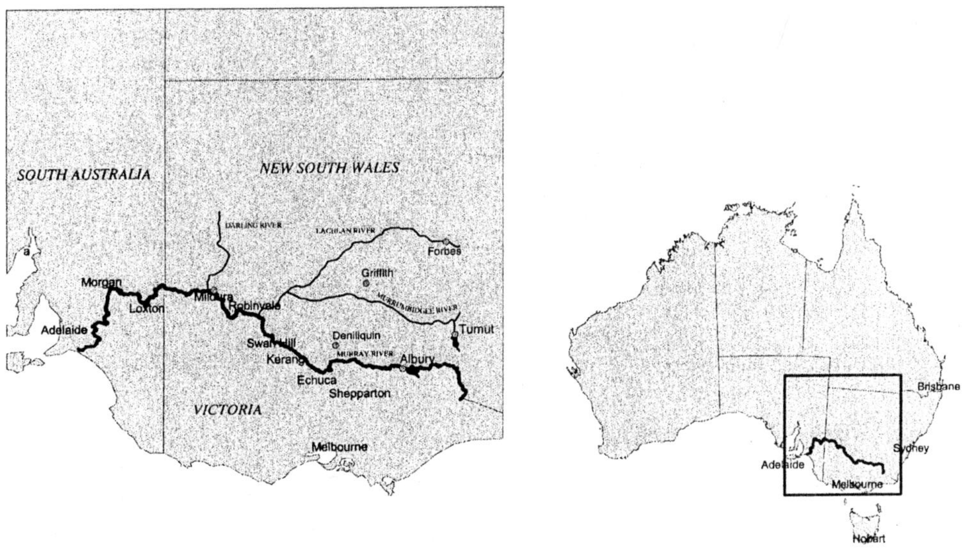

Figure 1. Major irrigation areas in the southern Murray Darling Basin

Table 1. Major irrigation areas in the southern Murray Darling Basin

Irrigation area	**Central town**
Goulburn–Broken	Shepparton
Campaspe	Echuca
New South Wales Murray	Deniliquin
Loddon – Barr Creek and Cohuna	Kerang
Loddon – Tragowel Plains	Kerang
Murrumbidgee	Griffith
Colignan	Robinvale
Mildura	Mildura
South Australian border to Lock 5	Loxton
Lock 4 to Lock 3	Loxton
Lock 3 to Lock 2	Loxton

2. BACKGROUND

The development of irrigation in the Murray Darling Basin was supported by public investment in infrastructure that began in the early 1900s, predominantly in the southern part of the basin along the Murray River and its tributaries. Two objectives of this investment were to increase agricultural exports and to move people back to rural Australia. This infrastructure has supported a large amount of low returning irrigated activities with low rates of irrigation efficiency. As irrigation water allocations were initially tied to the land, this inefficiency was preserved as these restrictions prevented water being redirected to its highest valued use. However, this did not limit irrigation development with the diversion of water from river systems in the Murray Darling Basin rising dramatically between the 1950s and mid 1990s. By the mid 1990s surface water use in the southern Murray Darling Basin had increased to about 6,000 GL with an estimated net value of irrigated production of more than one billion Australian dollars per annum (Hall et al., 1994). Total surface water use in the basin is estimated to be around 7,600 GL. Groundwater is also important in the basin, accounting for over 600 GL of consumptive water use; however these resources are distributed unevenly and vary substantially in terms of water quality (MDBC, 2001).

In 1995, an audit of water use in the basin showed that if the volume of water diversions continued to increase, this would exacerbate river health problems, reduce the security of water supply for existing irrigators in the basin, and reduce the reliability of water supply during long droughts. Consequently, a cap was imposed on the volume of water that could be diverted from the rivers for consumptive uses. While this cap limits further increases in water diversions, it does not constrain new developments provided the water for them is obtained by using current allocations more efficiently or by purchasing water from existing developments.

The cap on diversions has effectively made water, as opposed to storage and delivery infrastructure, a scarce resource and, thus, the need to develop an effective water market arose. However, there have been a number of impediments to the formation of a fully functioning water market and to date; there has not been a substantial change in water use. For example, within the Goulburn–Murray region temporary and permanent trade accounted for around five per cent of total allocations in 1996-97 (Earl and Flett, 1998). Of the 10 960 gigalitres of water diverted for irrigation in the Murray Darling Basin in 2001-02, only 10 per cent was traded (MDBC, 2003). The bulk of this trade was intraregional trade — that is, trade within valleys. Water trade between irrigation regions has been very limited, with some districts refusing to allow water to be traded out (Goesch and Beare, 2004).

Most of the impediments to trade can be attributed to the problem of changing from a centrally administered system of water allocations to a set of privately held water rights. These include the equity issues associated with allocating rights, and the sovereign risks associated with those rights. Beare and Bell (1998) demonstrated that access rights to infrastructure are also an important aspect of a water right when the timing of delivery is constrained by the capacity of the delivery system. It has also been noted that failure to link water rights to the fixed costs associated with infrastructure can lead to the stranding of irrigation assets (Goesch, 2001). Hence, while trade within irrigation areas is becoming established, there has been reluctance to allow trade between regions that do not share a substantial proportion of their delivery infrastructure. Further, to date the property rights issues associated with return flows from irrigation have not been considered. Return flows consist of surface runoff from flood irrigation, irrigation drainage and groundwater discharge from irrigation areas that reach the Murray River system.

Water trade and improvements in irrigation efficiency affect return flows that, in turn, have an impact on the quantity and quality of water used downstream. Due to the large volume of water that is diverted from the Murray River and its tributaries in the upstream irrigation areas and relatively low rates of irrigation efficiency, return flows form a substantial part of water available for downstream users. However, the salinity of return flows varies significantly between irrigation areas. There are two sources of irrigation salinity. The first is drainage and leaching of salts accumulated in the root zone that are discharged into the river system. The second source of irrigation salinity is the additional recharge and eventual increase in discharge from naturally saline groundwater systems. Soils throughout most of the irrigation areas in the southern Murray Darling Basin are shallow and the percolation of irrigation water through the soil has led to a large increase in the rate of recharge into the ground water system. Naturally saline ground water systems are the major source of regional differences in the quality of return flow as groundwater ranges from being relatively fresh in the upper catchments of the Murray River system to concentrations approaching seawater (more than 30,000 mg/L) at the lower end of the system. The associated increases in the levels of stream salinity in the Murray River can be seen in figure 2, where the irrigation areas under investigation are shown along the horizontal axis.

Return flows from irrigation areas with relatively low underlying groundwater salt concentrations may provide dilution flows downstream. Conversely, a reduction in return flows from upstream irrigation areas may increase the salinity of water supplies, imposing costs on downstream users. Irrigators presently hold an implicit right to return flows in that they can trade or save water without consideration of the downstream externalities. Undertaking these actions without explicit recognition of the downstream impacts generates externalities and leads to an inefficient allocation of water and suboptimal investment in improving irrigation efficiency.

These externalities have implications for both consumptive uses and environmental flows. In terms of consumptive use, rights to return flows are an equity, as opposed to an efficiency issue, so long as these rights are well defined. However, as irrigators are not required to account for a reduction in the volume of return flows as a result of their actions, these reductions are simply absorbed as an additional diversion imposed above the cap, which may be at the expense of desired environmental flows. Hence, the balance between rights to consumptive use and environmental flows needs to be specified with explicit consideration given to the rights to return flows.

However, a significant efficiency issue exists with the impact of return flows on water quality, in particular, the salt concentration of surface water flows. While the naturally occurring salinity of groundwater underlying the irrigation areas is the major determinant of the salt concentration of return flow, the level of salt discharge depends on groundwater recharge rates. This in turn depends on the area under irrigation, application rates, soil permeability and the level of irrigation efficiency. Hence the choice of location and irrigation practices can impose costs or generate benefits to downstream users that are not internalised in the costs and returns of those making the decision.

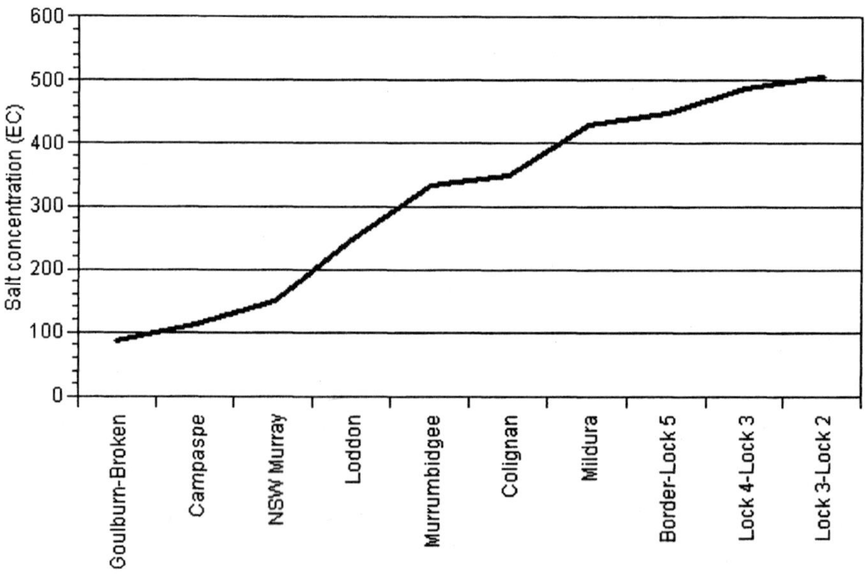

Figure 2. Salt concentration of the Murray River at tributary confluences, 2000

Water trade may also have an impact on river salinity as it alters the pattern of return flows from irrigation. For example, trade that moves water from an irrigation area with relatively low recharge rates and low groundwater salinity to a downstream irrigation area with high recharge rates and high groundwater salinity can produce a series of impacts on water quality. Immediately downstream of the seller, the transfer may increase stream flows and reduce salt concentration in the Murray River. However, as recharge rates are higher in the downstream area, surface runoff will be lower reducing the volume of return flows available downstream of the buyer. Further, as groundwater salinity is higher downstream, salt concentrations will be increased as more salt is transported to the river system. The impact of an investment that increases irrigation efficiency can also produce complex impacts on water quality. This is discussed in more detail in the results section.

The transaction costs associated with trying to fully internalise all the downstream impacts may be prohibitively high. Nevertheless, establishing site specific conditions on property rights associated with return flows may still lead to an improvement in economic welfare. To determine the potential magnitude of benefits of establishing such rights, SALSA – a catchment scale model – was developed to estimate the direct and downstream impacts of changes in irrigation practices.

3. MODEL SPECIFICATION

Within the modeling framework, economic models of land use are integrated with a representation of hydrological processes in each catchment. The unique feature of the model is linking of independent, as opposed to centrally planned, land and water use decisions with a full representation of the hydrological cycle. The hydrological component incorporates the relationships between rainfall, evapotranspiration and surface water runoff, the effect of land use change on groundwater recharge and discharge rates, and the processes governing salt accumulation in streams and soil. The interactions between precipitation, vegetation cover, surface water flows, groundwater processes and agricultural production are modeled at a river reach scale. In turn, these reaches are linked through surface and groundwater flows.

In the agro-economic component of the model, land use is allocated to maximise economic return from the use of agricultural land and irrigation water. Incorporated in this component is the relationship between yield loss and salinity for each agricultural activity. Thus, land use can shift with changes in the availability and quality of both land and water resources. Each land use area is optimised independently without taking into account the imposition of any downstream costs or benefits. This approach, initially proposed by Quiggan (1998), explicitly represents the generation of externalities associated with return flows. The theoretical framework underlying the modeling approach is described in Bell and Klijn (2000). The development and specification of the simulation model is discussed in further detail in Bell and Heaney (2000).

The rate at which salt stored in groundwater is transported to the river system is dependent on, among other things, the size of an irrigation development, irrigation application efficiency, the underlying geology of the irrigated area, and the distance between the irrigation development and the river valley. The methodology developed to assess the impact of changes in these parameters on salt loads in South Australian irrigation developments (Watkins and Waclawik, 1996; AWE, 2000) has been adapted to catchments in Victoria and New South Wales. Two specific changes were made. First, drainage schemes that discharge into the river system in many irrigation areas were incorporated into the model. In general, flows from these drains carry surface water runoff from flood irrigation and groundwater discharge. Second, the Murray River meanders in the Victorian Mallee (between the confluence of the Murrumbidgee and the South Australian border shown in map 1). As a consequence, groundwater may be either flowing toward or away from the river which, in turn, affects the level of saline stream discharge. This was incorporated by allowing a fraction of the recharge to move into a deep aquifer that does not discharge into the Murray River.

As the clearance of native vegetation has contributed to increased recharge in the dryland agricultural areas in the upland reaches of the catchments, the model also has land management units for rain-fed activities. However, as these areas are not affected by irrigation, they will not be considered here.

3.1 Agro-economic component

The management problem considered is that of maximising the economic return from the use of agricultural land by choosing between alternative steady state land use activities in each year. The model is static in the sense the optimisation does not anticipate the impact of irrigation on water tables and salinity levels, within or from outside a given irrigation area. However, as these resource conditions change, each region within the model is re-optimised. For tractability, land use in each irrigation area is classified into five activities, j, specified: irrigated crops, irrigated pasture, and irrigated horticulture, dryland crops and dryland pasture. The composition of these activities, and hence the relationships between land use, application rates and production vary between each region.

Each region is assumed to allocate its available land each year between the above activities to maximise the net return from the use of the land in production, subject to constraints on the overall availability of irrigation water from rivers sw^* and from groundwater sources gw^* and suitable land L^*:

$$\max \frac{1}{r} \sum_j p_j x_j \left(L_j, sw_j, gw_j \right) - csw \sum_j sw_j - cgw \sum_j gw_j \tag{1}$$

subject to

$$\sum_j sw_j \leq sw^*, \sum_j gw_j \leq gw^* and \sum_j L_j \leq L^* \tag{2}$$

where x_j is output of activity j. The decision variables are L_j, land use, sw_j, surface water use, and gw_j is groundwater use for irrigation of activity j. Static variables are r, the discount rate, csw the unit cost of delivery for surface water, cgw, the unit cost of extraction for groundwater and the net return to output for each activity. The net return is given by p_j and is defined as the revenue from output less the cost of inputs, other than land and water, per unit of output. Land use is measured in hectares and water use in megalitres or gigalitres.

For each activity, the volume of output depends on land and water use (or on a subset of these inputs) according to a Cobb-Douglas production function:

$$x_j = \begin{cases} A_j L_j^{\alpha_{Lj}} sw_j^{\alpha_{swj}(t)} gw_j^{\alpha_{gwj}} & 0 < \alpha_{Lj} + \alpha_{swj} + \alpha_{gwj} < 1 \quad for \quad j = 1,2,3 \\ A_j L_j^{\alpha_{Lj}} & 0 < \alpha_{Lj} < 1 \quad for \quad j = 4,5 \end{cases} \tag{3}$$

where A_j, α_{Lj}, α_{swj} and α_{gwj} are technical coefficients in the production function. Note, the technical coefficients on surface irrigation water are time dependent to capture the impact of changes in salt concentration in the Murray River.

The costs to irrigated agriculture and horticulture resulting from yield reductions caused by increased river salinity are modeled explicitly. The impact of saline water on the productivity of plants is assumed to occur by the extraction by plants of saline water from the soil. The electro-conductivity of the soil, EC, reflects the concentration of salt in the soil water and reduces the level of output per unit of land input (land yield) and per unit of water input (water yield). This is represented by modifying the appropriate technical coefficient, α_{swj}, in the production function for each activity from the level of those coefficients in the absence of salinity impacts, that is:

$$\alpha_{swj}(t) = \frac{\alpha_{swj}^{\max}}{1 + \exp\left(\mu_{0j} + \mu_{1j} EC\right)} \tag{4}$$

where μ_0 and μ_1 are productivity impact coefficients determined for each activity and α_{swj}^{max} is the level of the technical coefficients in the absence of salinity.

3.2 Hydrological component

There are two parts to the hydrological component of the model. The first is the distribution of precipitation and irrigation water between evapotranspiration (ET), surface water runoff and groundwater recharge. Evapotranspiration is determined as a function of precipitation and groundcover, as well as irrigation application rates and efficiency. Water application rates in the southern Murray Darling Basin for horticulture are around 10 megalitres per hectare a year, equivalent to 1000 mm of precipitation whereas average application rates for pasture are between 4 and 6 megalitres per hectare a year (Gordon et al., 2000). Irrigation efficiency is defined as the proportion of irrigation water applied that is returned to the atmosphere through ET. In horticultural areas such as Western Victoria and the South Australian Riverland, irrigation efficiency ranges between 75 and 80 per cent for horticulture (Anthony Meisner, Department of Environment, Heritage and Aboriginal Affairs, pers com, November, 2000). In areas where there is widespread use of flood irrigation on pasture, irrigation efficiency can approach 50 per cent.

The excess, precipitation and irrigation water less ET, is split between surface water runoff and groundwater recharge using a constant proportion (recharge fraction). The volume of irrigation water entering the groundwater system depends largely on terrain and soil structure. Irrigation areas are generally located in flat terrain leading to relatively high recharge fractions. On heavier soils in the upland river catchments, recharge fractions are assumed to range from 50 to 60 per cent. On the sandier soils in the South Australian Riverland recharge fractions are 100 per cent.

Some soils have intervening layers of clay that impede drainage into the groundwater system. Tile drainage is used in these areas to avoid waterlogging. Tile drainage is represented in the model through a combination of an increase in irrigation efficiency where drainage is re-used or allowed to evaporate, or as a return flow to the river system. Saline groundwater discharge can be intercepted through groundwater pumping for subsequent disposal in evaporation ponds. In some irrigation areas, such as the South Australian Riverland, there is groundwater discharge to the flood plains, which is mobilised in flood events and does not contribute to the problem of high salt concentrations. Reductions in average saline discharge from these effects are accounted for in calculating river salt and water balances.

The second part of the hydrology component is the determination of groundwater discharge. The equilibrium response time of a groundwater flow system is the time it takes for a change in the rate of recharge to be fully reflected in a change in the rate of discharge. The equilibrium response time does not reflect the actual flow of water through the groundwater system but the transmission of water pressure. The response time increases rapidly with the lateral distance the water flows in areas such as the South Australian Riverland due to the flat terrain and resultant low hydrological pressure.

Assuming the contributions of recharge are additive and uncorrelated over time, it is possible to model gross discharge directly, thereby avoiding the need to explicitly model groundwater levels. In the approach adopted here, total discharge rate *D* in year *t* is a logistic function of a moving average of recharge rates in the current and earlier years according to:

$$D(t) = R(0) + \sum_{i=t-m}^{t} \frac{R(i) - R(i-1)}{1 + \exp\left[\left(v_{half} - i\right) / v_{slope}\right]} \qquad (5)$$

where *R(0)* is the initial equilibrium recharge rate, *m* is the number of terms included in the moving average calculation, and υ_{half} and υ_{slope} are the time response parameters. The moving average formulation allows the accumulated impacts of past land use change to be incorporated as well as to model prospective changes.

As the distance from the river increases, the time before a change in the level of recharge is fully reflected in the level of groundwater discharge increases substantially. Irrigation areas in Western Victoria and the South Australian Riverland were divided into three land use bands according to distance from the river. Typical response profiles for the three land use bands are shown in figure 3. Parameters for the groundwater response functions in these irrigation areas were obtained from Watkins and Waclawik (1996). Similar groundwater response functions were assumed for the remaining irrigation areas based on discussions with CSIRO and other hydrologists. Response times were assumed to be longer the larger the irrigation area. However, in areas with substantial areas of high water tables, response times were reduced.

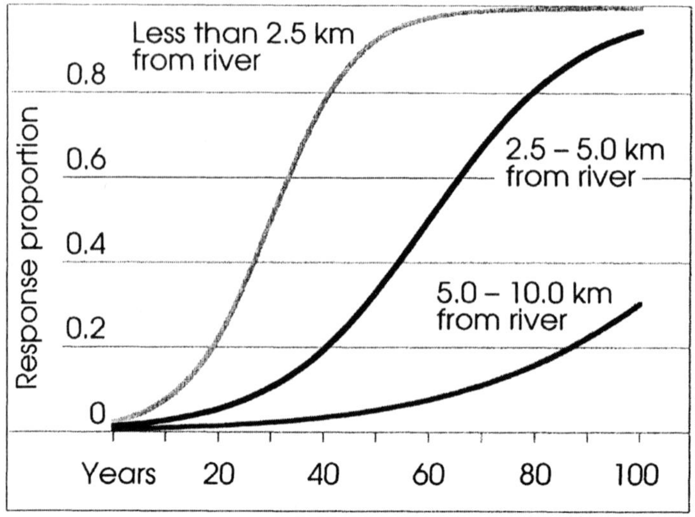

Figure 3. Weighting function for contribution of past recharge to discharge

3.3 Model calibration

The data required to calibrate the model are extensive. The procedure is presented in detail in Bell and Heaney (2000). Summary data for the irrigation areas is provided in Table 2. Additional information is available from the authors on request. Historical flows and salt loads were obtained from Jolly et al. (1997). Projected salt loads were obtained from the national salinity audit (MDBMC, 1999), Barnett et al. (2000) and Queensland Department of Natural Resources (QDNR, 2001). Land use and irrigation data was obtained from wide range of sources, including ABARE farm survey data and regional water authorities such as Goulburn–Murray Water and South Australia Water.

To calculate initial values for the production function parameters in (3), the total rent at full equity accruing to each activity was first calculated as the summation of rent associated with use of land and other fixed inputs to production and surface water. That is:

$$RentTotal_j = RentL_j + RentSW_j + RentGW_j + RentOther_j \qquad (6)$$

where

$$\begin{aligned} RentL_j &= L_j(0)\,p_{\min} \\ RentSW_j &= sw_j(0)\,c\tilde{s}w \\ RentGW_j &= gw_j(0)\,c\tilde{g}w \\ RentOther_j &= L_j(0)\left(p_j - p_{\min}\right) \end{aligned} \tag{7}$$

where p_{min} is the net return to land and other fixed capital structures in their marginal use and $c\tilde{s}w$ is the opportunity cost of surface water for irrigation and $c\tilde{g}w$ is the opportunity cost of ground water for irrigation in the initial period. Not all regions have groundwater sources suitable for irrigation. The opportunity cost of surface and ground water used for irrigation is assumed to be $50/ML for areas with predominantly pasture production and $200/ML for horticultural areas.

Table 2. Summary data for irrigation areas in the southern Murray Darling Basin, Australia

Irrigation area	Main irrigated activities	Water allocation		ET[a] fraction %	Recharge fraction[b] %	Groundwater salinity mg/L
		Murray GL	Tributary GL			
Goulburn–Broken	Pasture, cropping and horticulture	320	853	65	50	1000
Campaspe	Pasture and cropping	207	75	50	60	5000
New South Wales Murray	Pasture and cropping	2464	0	65	75	2000
Loddon Barr Creek, Cohuna	Pasture and cropping	371	30	65	75	2000
Loddon Tragowel	Pasture and cropping	455	0	55	75	9725
Murrumbidgee	Pasture, cropping and horticulture	0	2045	65	80	1000
Colignan	Horticulture	59	0	80	100	10000
Mildura	Horticulture	188	0	80	100	25000
Border-Lock 5	Horticulture	85	0	80	100	25000
Lock 4-Lock 3	Horticulture	93	0	80	100	21000
Lock 3-Lock 2	Horticulture	71	0	80	100	33000

a the percentage of irrigation was lost to evapotranspiration.

b the percentage of excess water, irrigation water and precipitation less evapotranspiration, that enters the groundwater system.

Initial values for the production function coefficients for each activity were then determined as:

$$\alpha_{Lj}(0)=\frac{RentL_j}{RentTotal_j}$$
$$\alpha_{swj}(0)=\frac{RentSW_j}{RentTotal_j}$$
$$\alpha_{gwj}(0)=\frac{RentGW_j}{RentTotal_j} \quad (8)$$
$$A_j=L_j(0)^{1-\alpha_{Lj}(0)}\,sw_j(0)^{-\alpha_{swj}(0)}\,gw_j(0)^{-\alpha_{gwj}(0)}$$

Within a simulation, these coefficients are then adjusted from the initial values according to equation (4). The coefficients in equation (4) were derived from estimated yield losses caused by irrigation salinity (MDBC, 1999) by equating the decline in average physical product of irrigation water with the yield loss function.

The Murray Darling Basin Commission has linked its hydrological modeling to estimates based on cost impacts of incremental increases in salinity. Costs downstream of Morgan are imputed as a function of EC changes in salt concentration at Morgan. The analysis considers agricultural, domestic and industrial water uses. Using the cost functions derived in this model, each unit increase in EC at Morgan is imputed to have a downstream cost of $65,000 (MDBC, 1999). This cost is included in the analysis presented here.

4. SIMULATION DESIGN

The model was initially used to determine a baseline over a 50 year simulation period. The estimated cost of salinity in the baseline scenario is measured as the reduction in economic returns from agricultural and horticultural activities from those that are currently earned. Thus, only costs and/or benefits associated with changes in stream flows, salt concentration and the extent of high water tables from current levels are estimated. Salt loads and salt concentration of the Murray River are predicted to rise over the next 50 years as a result of both the clearance of native vegetation to facilitate dryland agriculture and the increased mobilisation of salt associated with irrigated agriculture. The salt concentration at Morgan, a gauging site on the Murray River below the major irrigation areas, is projected to increase from 567 EC currently to 650 EC by 2050. This increase in salt concentration is expected to result in a decline in agricultural returns of almost $300 million, in net present value terms (NPV) using a discount rate of five per cent, and impose costs to agricultural, urban and industrial water users downstream of Morgan of $42 million NPV over the 50 year period.

Two series of simulations were conducted for each of the major irrigation areas on the Murray River system to allow a comparison of the internal and external costs or benefits of changes in irrigation allocations and practices relative to the baseline scenario. The irrigation areas under consideration, listed in Table 1, are shown in upstream to downstream order in figure 1. Internal impacts are derived within the irrigation undertaking the action whereas external impacts are those derived downstream of the area undertaking the action. In the first series, water allocations were reduced by 20 GL in each irrigation area. These reductions were sourced from the Murray River as opposed to the tributary rivers. The internal and external costs and benefits associated with a reduction in water allocations and return flows were then calculated over a 50 year time period.

In the second series of simulations, irrigation efficiency was increased by five per cent. With the increase in efficiency the fraction of irrigation water applied which returns to the atmosphere increases by 5 per cent. At the same time, an equivalent percentage reduction in the volume of irrigation water applied results in the same crop yield. It was assumed that irrigators retain all water savings and use those savings to expand irrigated production. Hence the reduction in surface water and groundwater recharge will be less than five per cent, depending on the absolute level of irrigation efficiency. This series illustrates the impact of changes in volumes of water available for irrigation, as downstream allocations are determined as shares of available flows. Again, the internal and external benefits and costs were calculated over a 50 year time period.

5. RESULTS

5.1 Reduction in water allocation

The external impacts of a reduction in allocation on water quality arise from two sources that may produce either benefits or costs at different locations along the river system. First, as the water that would have otherwise been used for irrigation is retained in the river, there is an immediate reduction in salt concentration. Second, a reduction in groundwater recharge due to the reduction in irrigation allocation results in reduced discharge from the groundwater system and will lower salt loads over time. The effect on salt concentration of the Murray River will depend on the difference in salt concentration between groundwater and stream flows at different points along the river.

The internal costs and the external benefits derived from 20 GL reduction in irrigation water allocations are shown for each irrigation area in figure 4. The internal costs are a result of forgone irrigated production with the highest costs incurred in the areas dominated by high value horticulture.

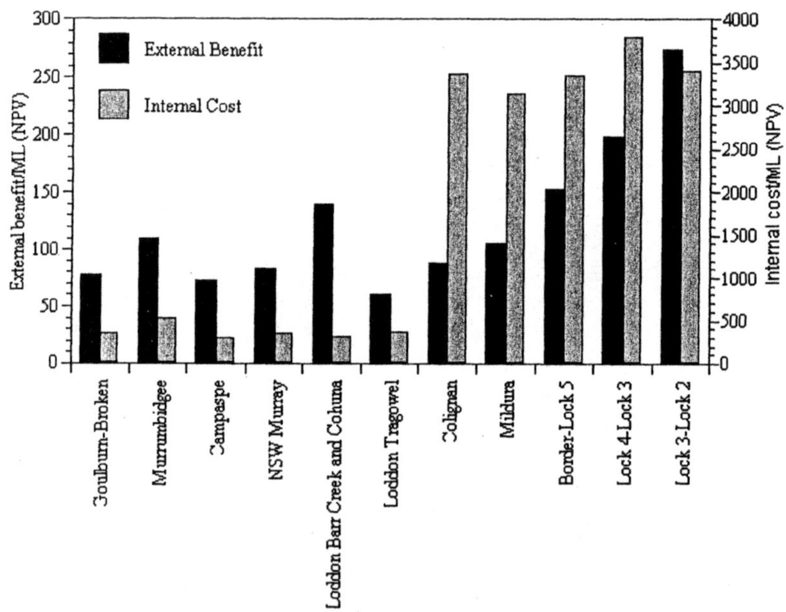

Figure 4. External benefits and internal costs per megalitre of a 20 GL reduction in water allocations

The external benefits from a reduction in water allocation vary substantially between irrigation areas. In the upper catchments of Victoria and New South Wales where recharge is high due to low rates of irrigation efficiency, the external benefits are high relative to the value of water use despite low levels of groundwater salinity. This, in part, reflects the location of these irrigation areas in the upper reaches of the river system and the predominance of low value irrigated agriculture.

In contrast, in the South Australian Riverland and Western Victoria, the external costs of water use are large due to high levels of ground water salinity. The reduction in the groundwater discharge component of irrigation return flows reduces the volume of salt transported to the river and improves the quality of water for downstream uses. However, the internal costs incurred as a result of forgone irrigated activity are also high as these regions are dominated by high value horticultural production.

These findings are consistent with findings of Weiberg et al. (1993) who highlighted the potential importance of recognising the positive externalities associated with trade. This work demonstrates that the impacts of trade in water between irrigation areas can generate both external benefits and costs that are significant. If water from the Goulburn–Broken was traded to the reach between Lock 3 and Lock 2, for example, the cost of forgone agricultural production in the Goulburn–Broken would be around $355/ML. The net effect of trade would be to increase the external costs of irrigation by almost $200/ML, from around $75/ML in the Goulburn–Broken to around $275/ML between Lock 3 and Lock 2. To put this into perspective, the price of a permanent water entitlement in a South Australian irrigation area was around $500/ML (Samaranayaka et al., 1998). To fully account for the externality associated with trading from the example above, the price of permanent water allocation would need to increase by 40 per cent to around $700/ML.

In contrast, while trading water upstream from the Loddon Bar Creek and Cohuna irrigation areas to the New South Wales Murray, for example would not substantially alter the agricultural returns to irrigation, it would generate an external benefit. This benefit arises because the externality associated with irrigation in the Loddon Barr Creek and Cohuna area is higher than that associated with the New South Wales Murray. The net reduction in the external cost of irrigation as a result of upstream trade between these two areas would be around $60/ML.

However, in both of the examples above, the individuals who trade do not accrue all benefits and costs associated with a change in water quality. Hence, even if there was a property right associated with the physical change in return flows at the source and destination, its traded price would not reflect its full value.

5.2 Improvements in irrigation efficiency

Internal benefits from increased irrigation efficiency are derived from an increase in agricultural revenue as a result of increased availability of irrigation water (figure 5). External salinity benefits from improvements in irrigation efficiency are derived from reductions in the discharge of saline water directly into streams, which leads to a reduction in the salt load and concentration of river flows and hence, an improvement in the quality of water available for downstream users. The extent to which a reduction in salt loads and concentration is achieved depends on, among other things, the volume of the reduction in recharge and the underlying groundwater salinity. As a result of the improvement in water quality, agricultural yields and revenue increase. The main driver of the benefit profile is the response time of the groundwater aquifer with shorter response times generating water quality benefits sooner. External benefits are only derived as a result of improvements in irrigation efficiency in the lower reaches of the Murray River system where groundwater salt concentrations are high and groundwater response times are short relative to those in the upper reaches of the system (figure 5).

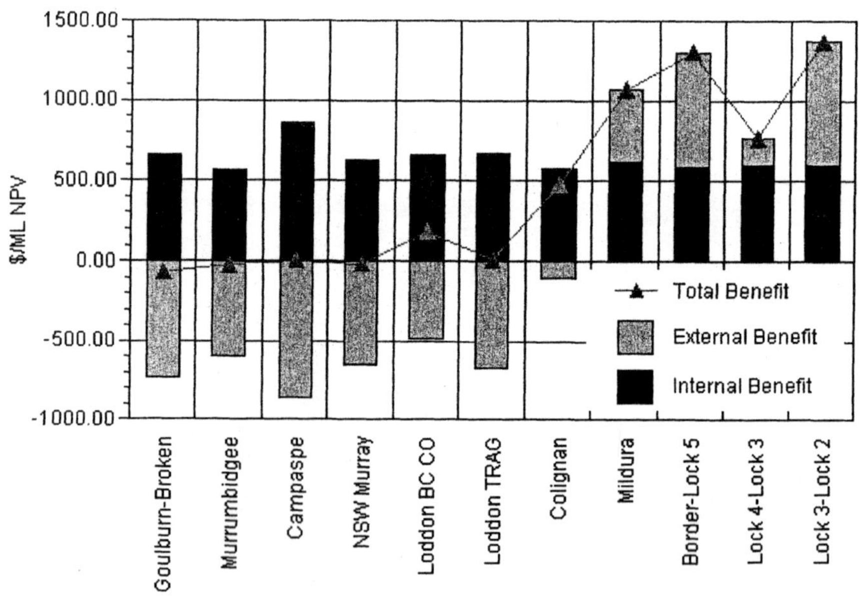

Figure 5. External benefits and costs per megalitre of a 5 per cent increase in irrigation efficiency

An improvement in irrigation efficiency in the upper catchments generates an external cost. As these areas are characterised by large volumes of surface water runoff and low groundwater salt concentrations, the reduction in return flows from irrigation increases salt concentration in the Murray River reducing the productivity of irrigation water in downstream uses. Further, under conditions where total extractive use is capped, the reduction in return flows reduces the quantity of irrigation water available for use in downstream irrigation areas.

6. CONCLUDING REMARKS

Water trade and increased irrigation efficiency can affect return flows and have subsequent impacts on downstream allocations and water quality. These downstream impacts of changes in return flows are very diffuse and, in the case of water quality, generate both positive and negative externalities at different points along the river system. As those who engage in water trade and invest in improved irrigation efficiency do not bear the external costs or benefits of their actions, the level of action undertaken is likely to be sub-optimal from the combined perspective of all water users.

Irrigators presently have an implicit right to the return flows in that they can trade or save water without consideration of any downstream externalities their actions generate. To achieve an economically efficient level of trade or investment in irrigation efficiency, the downstream costs or benefits associated with changes in return flows need to be internalised into the decisions faced by upstream irrigators. Because these impacts are very diffuse, the transaction costs associated with establishing property rights that fully internalise the effect of return flows on downstream users are likely to be prohibitive. However, there may be potential economic gains from attaching site specific conditions to the implicit rights to return flows. In the case of trade, this may take the form of charges or subsidies attached to trade that lead to higher or lower costs associated with downstream changes in river salinity.

In the case of improved irrigation efficiency, investment incentives should reflect the net downstream impacts of reduced return flows. Irrigators' rights to water saved through improved efficiency will influence their incentives to adopt or invest in water saving practices and technology. The nature of these rights would need to be location specific to have an economically efficient level of investment in improving irrigation efficiency. For example, in the upper catchments where there are negative externalities due to reduced return flows, irrigators may be entitled to retain a proportion of the water saved. In the lower reaches of the Murray River where there are positive externalities associated with reduced return flows, irrigators may need to receive compensation in excess of their water savings to generate an efficient level of investment.

Griffin and Hsu (1993) noted that the instream or environmental benefits or costs may also be an important issue that can not be addressed by simply defining property rights over the quantity and quality of return flows. Institutional arrangements are required to create appropriate economic incentives to ensure that investments in increased water use efficiency and water trade lead to the best outcome for society as a whole. With increasing concerns for riverine habitats in the Murray Darling Basin, this is key issue for ongoing research.

Notes

1. The most widely used method of estimating the salinity concentration in water is by electrical conductivity. To convert 1 EC to mg/L total dissolved salts, a conversion factor of 0.6 generally applies.

References

AWE (Australian Water Environments), 1999. Assessment of the Trends in Salt Loads and Salinity along the River Murray in South Australia, Final Report, May.

Barnett, S., W. Yan, N.L. Watkins, J.A. Woods and J.M. Hyde, 2000. Murray Darling Basin Salinity Audit – Groundwater Modelling to Predict Future Salt Loads to the River Murray in SA, Department for Water Resources, South Australia, Adelaide, December.

Beare, S. and R. Bell, 1998. 'Efficiency of water use and access rights to infrastructure', *Australian Commodities,* 5(4), 494–503.

Bell, R. and A. Heaney, 2000. 'A basin scale model for assessing salinity management options: model documentation', ABARE technical working paper 2000.1 (www.abareconomics.com), Canberra, August.

Bell, R. and N. Klijn, 2000. 'A basin scale model for assessing salinity management options', ABARE paper presented at the *Xth World Water Congress*, Melbourne, 11–17 March.

Earl, G. and D. Flett, 1998. 'Water trading in Victoria's Goulburn–Murray regions – a status of practice vs theory', in *Outlook '98*, proceedings of the National Agricultural and Resources Outlook Conference, Canberra, 4–6 February, vol. 1, *Commodity Markets and Resource Management*, ABARE, Canberra, pp.150–153.

Goesch, T., 2001. 'Delivery charges for water: their impact on interregional trade in water rights', *Australian Commodities,* 8(4), 626–634.

Goesch, T. and S. Beare, 2004. 'Water rights and trade: meeting the water reform agenda', *Australian Commodities,* March 2004, pp.180–190, ABARE, Canberra.

Gordon, S., A. Kemp and C. Mues, 2000. *Irrigation in the Murray Darling Basin and the impact of water policy reforms,* ABARE report prepared for the Natural Heritage Trust MD 2001 – Fish Rehab Program, Canberra, August.

Griffin, R.C. and S.H. Hsu, 1993. 'The potential for water market efficiency when instream flows have value', *American Journal of Agricultural Economics,* 75(2), 292–303.

Hall, N., D. Poulter and R. Curtotti, 1994. *ABARE Model of Irrigation Farming in the Southern Murray Darling Basin*, ABARE Research Report 94.4, Canberra, June.

Heaney, A., S. Beare and R. Bell, 2000. 'Targeting reforestation for salinity management', *Australian Commodities,* 7(3), 511–518, ABARE, Canberra.

Heaney, A., S. Beare and R. Bell, 2001. 'Evaluating improvements in irrigation efficiency as a salinity mitigation option in the South Australian Riverland', *Australian Journal of Agricultural and Resource Economics,* 45(3), 477–493.

Heaney, A., S. Beare and T. Goesch, 2002. 'Environmental flows and water trade', ABARE Current Issues 02.3, Canberra, March.

Jolly, I.D., T.I. Dowling, L. Zhang, D.R. Williamson and G.R. Walker, 1997. 'Water and salt balances of the catchments of the Murray-Darling Basin', *CSIRO Land and Water Technical Report 37/97*, Melbourne, November.

MDBC (Murray Darling Basin Commission), 1999. *Salinity Impact Study,* Report by Gutteridge Haskins & Davey Pty Ltd, Canberra, February.

MDBC (Murray Darling Basin Commission), 2001. *Natural Resource Management*, http://www.mdbc.gov.au, Canberra.

MDBC (Murray Darling Basin Commission), 2003. Water Audit Monitoring Report 2001-02, Canberra.

MDBMC (Murray–Darling Basin Ministerial Council), 1999. *The Salinity Audit of the Murray–Darling Basin: A 100 Year Perspective*, Canberra, October.

QDNR (Queensland Department of Natural Resources), 2001. *Projection of groundwater discharge and salt loads for the Queensland catchments of the Murray–Darling Basin – Draft report*, Resource Sciences & Knowledge, QDNR, Indooroopilly.

Quiggan, J., 1998. 'Murray River salinity – an illustrative model', *American Journal of Agricultural Economics,* 70, 635–645.

Samaranayaka, D., F. Freeman and C. Short, 1998. 'Water trading in the Murray–Darling Basin – some preliminary observations', in *Outlook '98*, proceedings of the National Agricultural and Resources Outlook Conference, Canberra, 4–6 February, vol. 1, *Commodity Markets and Resource Management*, ABARE, Canberra, pp. 150–153.

Watkins, N.C. and V.G. Waclawik, 1996. River Murray Water Resource Management Assessment of Salt Load Impacts and Drainage Hazard for New Irrigation Development Along the River Murray in South Australia, RD 96/17, Department of Mines and Energy, Adelaide.

Weinberg, M., C.L. Kling and J.E. Wilen, 1993. 'Water markets and water quality', *American Journal of Agricultural Economics,* 75(2), 278–291.

Chapter 10

ECONOMIC ANALYSIS OF GREEN PAYMENTS TO PROTECT WATER QUALITY

Madhu Khanna
University of Illinois at Urbana-Champaign, U.S.A.

Richard Farnsworth
Purdue University, U.S.A.

1. INTRODUCTION

Agricultural nonpoint source pollution caused by soil erosion and extensive use of chemical nutrients and pesticides is a major source of degradation of water quality in the United States. The control of nonpoint pollution remains one of the most difficult policy challenges. The diffuse nature of nonpoint emissions and their stochasticity and natural variability due to weather and other environmental processes makes it difficult to observe and measure them at reasonable cost. Several approaches to protect water quality from nonpoint pollution have been used by the US Department of Agriculture (USDA) and by the US Environmental Protection Agency (USEPA). These approaches include various forms of voluntary assistance, such as education, technical assistance and cost sharing to promote production practices that reduce the negative impact of agriculture on water quality as well as regulatory mechanisms that rely on design standards, performance standards and trading between point and nonpoint sources (USEPA, 2001). The 1985, 1996, and 2002 Farm Bills greatly expanded USDA's use of incentives and cost-share instruments to address agro-environmental degradation through programs such as the Conservation Reserve Program (CRP), the Conservation Reserve Enhancement Program (CREP) and the Environmental Quality Incentives Program (EQIP). These programs seek to provide financial incentives or "green payments" for farmers to change observable choices, such as land use (crop production or retiring the land and converting it to permanent grasses), level of input use, and technology (irrigation methods, crop rotation or tillage practices), thus making agriculture more environmentally friendly.

The reliance on green payment policies to achieve environmental goals raises several questions. What criteria should be used by a policy maker to select the recipients of these payments? How should these payments be designed to achieve a social planner's objectives in a decentralized decision making situation, where participation in a green payment program is a voluntary decision of the farmer? How efficient are green payment programs relative to first best instruments to control pollution that would have maximized social welfare?[1] This chapter reviews the recent methodological developments in the literature on these issues.

The generation of nonpoint source pollution and the environmental damages caused by it vary by location. Differences in farming practices, land quality, climate, topography, and hydrological characteristics that exist even in relatively small areas contribute to this heterogeneity across the landscape. Additionally, the costs of adopting alternative environmental practices also vary with land quality, climate and management skills. Thus the extent to which regulators should be willing to make green payments to achieve environmental goals and the payments needed to induce environmentally friendly activities can be expected to vary across locations.

Implementing a green payment program therefore requires a mechanism to select/target the recipients of green payments. A social planner could choose to maximize environmental benefits, minimize costs of changing practices, or combine the two objectives. Babcock et al. (1996, 1997) show how different selection/targeting rules produce different outcomes in terms of the land selected for retirement depending on the correlation between the distribution of environmental benefits and economic costs within a region as discussed in Section 2.1. The implementation of any of these selection rules, in the presence of heterogeneity in land and farmer characteristics, requires site-specific information about the environmental quality benefits of alternative choices and/or the costs of those choices to the farmer. The determination of environmental benefits involves linking production practices to on-site pollution generation and then linking the latter to off-site pollution loadings in water bodies. Khanna et al. (2003) and Yang et al. (2003) show the complexity in specifying such a relationship because it varies not only with location and other characteristics of a land parcel but it also depends on the decisions made on neighboring land parcels. In Section 2.2, we present their framework to examine methods that can be used to develop criteria for selecting land parcels that should be enrolled in a land retirement program.

Though a social planner can select targeting criteria and rules to identify land parcels to participate in green payment programs and levels of payments to achieve social efficiency, he/she cannot command participation by farmers. Instead participation needs to be voluntarily induced by providing adequate incentives. This implies that green payment instruments need to be designed to create incentives for voluntary participation selectively. This requires determining the appropriate level of green payments and also how these payments should vary across land parcels depending on their heterogeneous location and site-specific characteristics as discussed in Section 3.1. Since the decision to adopt an environmentally friendly technology has to often be made under uncertainty about yields, prices, weather and/or the performance of the technology, the level of green payments required to induce adoption may depend not only on the profits foregone due to adoption but also on risk attitudes of farmers and option values they attach to waiting for more information before making irreversible decisions that involve sunk costs. We discuss the methods used to analyze green payments under uncertainty in Sections 3.2 and 3.3.

Much of the literature on nonpoint pollution control policy has focused on analyzing the relative efficiency of alternative types of taxes, standards and tradable permit policies (Horan and Shortle, 2001). Relatively few studies have examined the implications of a green payments policy relative to an effluent tax for reducing water pollution (Wu and Babcock, 1999; Khanna et al., 2002). A pollution tax is appealing because it achieves abatement through a cost-effective mix of three mechanisms, a negative extensive margin effect (exit of polluting and less productive farms), a negative intensive margin effect (a reduction in input-use) and a technology switching effect (adoption of environmentally friendly technologies). Green payments in the form of uniform cost-share subsidies and input reduction subsidies are much more restricted in the types of incentives each one provides for reducing the use of polluting inputs and inducing the adoption of a conservation technology. Additionally, green payments also lead to deadweight losses associated with the government expenditures needed to finance the subsidies. However, an emissions tax policy involves enforcement and implementation costs. Studies examining conditions under which green payments may be more efficient than a benchmark emissions tax policy and the magnitude of welfare losses due to reliance on a uniform green payment policy are reviewed in Section 4.

In sum, this chapter presents the latest methods used to analyze three issues related to the use of green payments for protecting water quality from agricultural nonpoint pollution: the implications of alternative decision rules for targeting green payments by a policy maker, the design of green payment instruments in the presence of spatial heterogeneity, risk and uncertainty, and finally their efficiency relative to an emissions tax policy.

2. TARGETING OF GREEN PAYMENTS

2.1 Alternative Decision Rules for a Social Planner

One of the first issues a social planner must address is the selection of a decision rule for implementing a green payment program. Babcock et al. (1995; 1997) examine the implications of alternative rules for programs where farmers receive payments to remove cropland from production as a means of reducing damages caused by erosion and run-off. They consider three alternative decision rules for green payments for land retirement that can be applied subject to an aggregate budget constraint, TC^*. These are: maximize acreage enrolled in the program, maximize environmental benefits, and maximize the benefit to cost ratio.

Consider a region with per acre annual costs of land acquisition that range between $C_o \leq C \leq C_I$ and per acre annual benefits from land retirement that range between $B_o \leq B \leq B_I$. Costs vary because the land varies in its productivity and thus in the quasi-rents earned under crop production while environmental benefits may vary because of location relative to sensitive natural resources and soil conditions. The joint distribution of land available with benefits and costs within these ranges is shown by the encircled area in Panels A and B of Figure 1. In panel A, benefits and costs are negatively correlated while in panel B they are positively correlated. The social planner's problem is to select a subset of the C and B values while meeting the budget constraint. Maximization of acreage is accomplished by ranking land parcels from low to high cost and retiring land until the budget TC^* is exhausted. Suppose the highest cost land accepted under this rule is C^*. This implies that all land in areas D+G+H+I would be retired from crop production in Panel A and Panel B of Figure 1.

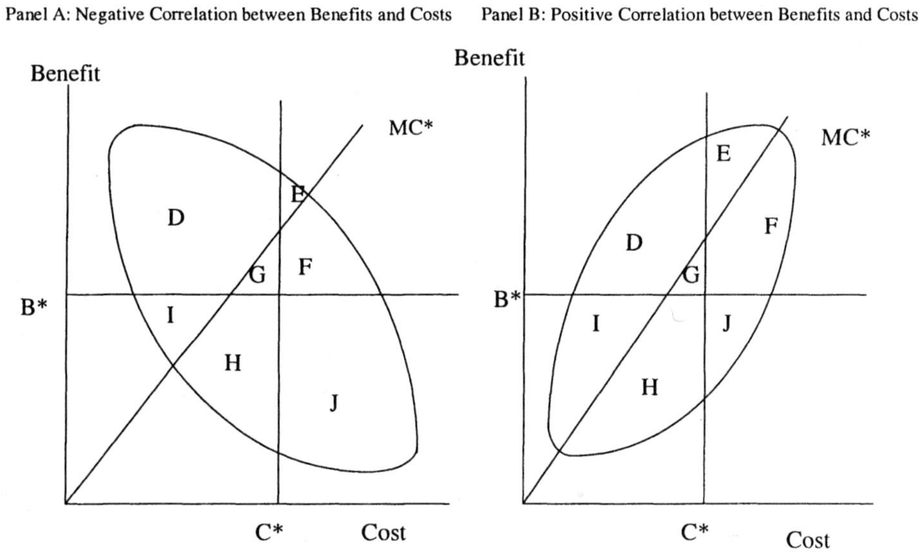

Fig. 1: Effects of Alternative Decision Rules For Green Payments on Land Targeted for Retirement

Alternatively, environmental benefits could be maximized by ranking land from high to low *B* and enrolling the most environmentally sensitive land first. Suppose that the smallest per-acre benefit accepted under this targeting scheme is B*. Land retired from production would lie in areas D+G+E+F. The third decision rule takes into account benefits and costs and ranks land from low to high according to the marginal cost of providing *B* which is measured by *C/B*. Let *MC** denote the highest marginal cost that can be enrolled given the budget, which implies that land in areas I+D+E will be enrolled. Babcock et al. (1997) show that the extent to which the alternative rules result in different outcomes depends upon the characteristics of *s(B,C),* represented by the encircled areas shown in Figure 1. With a negative correlation between *B* and *C* all three targeting schemes would select land in area D and none in area *J*. Thus outcomes from the three schemes would tend to converge. On the other hand, with a positive correlation between *B* and *C* as in panel b of Figure 1, environmental benefit targeting would enroll land in areas *E* and *F* while cost targeting would enroll land in areas *I* and *H*; two very disparate outcomes.

They also show that a mean preserving spread in the distribution of benefits, holding B* constant increases the level of benefits achieved under benefit targeting but does not affect the level of benefits under cost targeting. A mean preserving spread in the distribution of costs will increase C* and the amount of land that can be selected and the level of benefits achieved under cost-targeting. Thus, variability as well as correlation determine the difference in outcomes under cost and benefit targeting.

The joint distribution of costs and benefits determines the magnitude of the difference in outcomes under alternative targeting tools. With a normal distribution, an increase in positive correlation decreases the level of benefits that can be achieved for a given budget constraint under both cost and benefit targeting. An increase in cost variability increases the benefits that can be obtained from cost targeting and has no effect on the benefits that can be obtained from benefit targeting when correlation is zero. When the correlation coefficient is negative, an increase in cost variability increases the benefits obtained from benefit targeting.

2.2 Incorporating the pollution generation process in a decision rule

Babcock et al. (1996; 1997) demonstrate the importance of knowing the distribution of environmental benefits and costs and their implications for the land that would be selected under different decision rules in order to optimally target a land retirement program. However, quantification of the environmental benefits, even in physical terms, from retiring individual land parcels is complex. In the case of sediment and other chemical pollutants that contaminate water quality, it involves linking on-site generation of pollution with its off-site loadings in a water body through a pollution transport relationship.

Carpentier et al. (1998), Prato and Wu (1996), and Ribaudo (1989) include a fixed sediment transport process for estimating movement of sediment off a parcel of land, through downslope parcels and eventually to a stream or other water body. The link between the amount of sediment generated by a land parcel and the amount reaching a water body is assumed to be either a fixed proportion or dependent only on the distance of the parcel from the water body. However, the portion of soil transported from a land parcel to a water body depends on that parcel's site-specific characteristics (slope, soil characteristics, and distance from a water body) and land use decision (crops, trees, pasture or grass) as well as on the land use and sediment trapping coefficients of downslope land parcels. It also depends on the volume of run-off flowing in from upland parcels which depends on land use decisions and site-specific characteristics of upslope parcels. Therefore, the environmental contribution of retiring a land parcel cannot be determined exogenously simply based on its own on-site erosion and fixed site-specific characteristics of intervening land. There is instead a need to use more detailed spatial information about the location and other characteristics of land parcels and to determine the benefits provided by each parcel in a sediment flow path jointly or endogenously with the land use decisions of all parcels in that flow path. Lintner and Weersink (1999) incorporate the interdependence between sediment deposition coefficients and fertilizer-use decisions of all parcels in a flow path, but simplify the problem by assuming that all parcels are identical and thus make the same land use decisions.

Khanna et al. (2003) develop a framework to endogenously and simultaneously determine the sediment trapping efficiency and the land use decisions of all parcels along the same flow path. They examine the characteristics of the land parcels that should be targeted for enrollment in a land retirement program, such as CREP, to achieve given sediment abatement goals at least cost. Yang et al. (2003) extend this analysis to examine the criteria for cost-effective allocation of sediment abatement responsibility across watersheds. Results from these studies suggest the following framework for creating a least cost land retirement program with pre-determined abatement standards.

Suppose there are n=1, 2,..., N watersheds in a region. Each watershed contains j=1,...,J_n surface runoff channels , and each runoff channel consists of i=1,..I_{jn} homogeneous land parcels of equal size, say α. Pollutants move from the highest numbered parcel in a flow channel to the lowest numbered parcel (i=1), which borders the water body. Furthermore, runoff channels are independent of each other. Yang et al. (2003) assume that there is homogeneity of soil characteristics, erodibility and other processes within a parcel but heterogeneity across parcels. Each parcel chooses the amount of land X_{nijk} allocated to activity k where k=0 denotes land retirement and k=1 denotes crop production. Let π_{nijk} denote the per acre quasi-rent earned on the ith land parcel in the jth channel with the kth activity in watershed n. Quasi-rent equals revenue minus variable costs.

The on-site sediment generated per acre by the kth activity in the ith parcel is denoted by s_{nijk}, and total sediment produced by the ith parcel is $\sum_{k=0}^{1} s_{nijk} X_{nijk}$. Some of this sediment is deposited on downslope land parcels and does not cause any damage to off-site water quality. Let $d_{n,i,i-m,j}$ denote the fraction of the sediment originally produced by the ith parcel and deposited in each of the i-m downslope parcels in flow path j, where m=0,..,i-1. It also represents the deposition ratio of land parcel i-m. These deposition ratios are a function not only of the site-specific characteristics, L_{nij} and land use activities, X_{nijk} of each of the downslope parcels but also a function of the amount of sediment inflow, $S_{n,i+1,j}$, from upland parcels. Each parcel's deposition ratio is a function of its own land use and the land use of upslope and downslope land parcels and represented as follows:

$$d_{n,i,i-m,j} = d(L_{n,i-m,j}, X_{n,i-m,j,k}, S_{n,i-m,j}) \quad \text{for } m = 0,\ldots, i-1 \tag{1}$$

where:

$$S_{n,i-m,j} = s(L_{n,i-m+1,j},\ldots, L_{n,I_j,j}; X_{n,i-m+1,j,k},\ldots, X_{n,I_{jn},j,k}) \tag{2}$$

The fraction of sediment generated by the ith parcel and deposited in the flow channel and not loaded into the water body is $\sum_{m=0}^{i-1} d_{n,i,i-m,j} \le 1$.

We can now overlay the decision rule (minimize the costs of achieving a sediment abatement target $\bar{A}$) on this conceptual specification of the sediment generation process to examine the characteristics of the land parcels that should be retired from crop production. If S_n^0 is the sediment loading in watershed *n* before land retirement, the optimization problem is:

$$\text{Min} \sum_{n=1}^{N} \sum_{j=1}^{J_n} \sum_{i=1}^{I_{jn}} \pi_{nij1} X_{nij1} - \sum_{n=1}^{N} \sum_{j=1}^{J_n} \sum_{i=1}^{I_{jn}} \sum_{k=0}^{1} \pi_{nijk} X_{nijk} \tag{3}$$

subject to:

$$\sum_{k=0}^{1} X_{nijk} = \alpha, \ \forall n,i,j, \tag{4}$$

$$\sum_{n=1}^{N} S_n^0 - \sum_{n=1}^{N} \sum_{j=1}^{J_n} \sum_{i=1}^{I_{jn}} (1 - \sum_{m=0}^{i-1} d_{n,i,i-m,j}) \sum_{k=0}^{1} s_{nijk} X_{nijk} \ge \bar{A} \tag{5}$$

The objective function is the loss in profits from crop production due to land retirement. These are zero when *k=1*. From the first order optimality conditions, land retirement on parcel *i* is socially preferable if:

$$\lambda^* [(1 - \sum_{m=0}^{i-1} d^*_{n,i,i-m,j})(s_{nij1} - s_{nij0}) + 2 s_{nij0} \alpha \sum_{m=0}^{i-1} \frac{\partial d_{n,i,i-m,j}}{\partial X_{nij0}} + 2 s_{nij0} \alpha \sum_{m=1}^{I_{jn}-i} \frac{\partial d_{n,i+m,i,j}}{\partial X_{nij0}}] > \pi_{nij1} \tag{6}$$

where λ^* is the Kuhn-Tucker multiplier (shadow price) associated with constraint (5) and represents the marginal value per ton of the abatement achieved by retiring a land parcel. It is the amount a planner would be willing to pay per ton of abatement to induce voluntary land retirement by farmers to achieve the sediment abatement target $\overline{A}$. Furthermore, this marginal value per ton is same for all watersheds, implying that a ton of abatement from any watershed should be valued equally. If the program's region consists of watersheds with significantly different topology, production practices and other characteristics, more abatement will occur in watersheds that exhibit low abatement costs compared to watersheds with higher abatement costs. Note that due to the homogeneity assumption, the optimal solution consists of parcels that are either fully retired or fully under crop production.

The larger the three terms on the left hand side and the smaller the term on the right hand side of (6), the greater the net benefits from retiring the parcel. A closer examination of the three terms on the left hand side of (6) is useful for understanding the types of parcels favored in a green payment sediment reduction program. The term $(1-\sum_{m=0}^{i-1} d^*_{n,i,i-m,j})(s_{nij1}-s_{nij0})$ captures the off-site abatement of sediment generated on parcel i due to a change in its land use. This positive term can increase two ways. A large positive value for $(s_{nij1}-s_{nij0})$ implies that land retirement significantly reduces a parcel's sediment production, a desirable consequence given a sediment goal. Second, a larger portion of sediment generated is loaded in a water body as $\sum_{m=0}^{i-1} d^*_{n,i,i-m,j}$ decreases. The term $\sum_{m=0}^{i-1} d^*_{n,i,i-m,j}$ tends to be small if the ith parcel is adjacent to the stream or nearby, the downslope parcels do not trap much of the sediment generated by the ith parcel, or the amount of sediment flowing into the ith parcel from upslope parcels in the flow chain is large.

The second term, $\sum_{m=0}^{i-1}(\frac{\partial d_{n,i,i-m,j}}{\partial X_{nij0}}) > 0$, represents the positive impact that a land use change on the ith parcel has on the deposition ratios of the $(i\text{-}m)$ downslope parcels. The indirect benefit of land retirement by the ith parcel is large if it significantly raises deposition ratios of down-stream parcels. These benefits through trapping sediment flows on downslope parcels become particularly important if the volume of sediment generated by the ith parcel even after land retirement is large.

The third term, $\sum_{m=1}^{I_{jn}-i}\frac{\partial d_{n,i+m,i,j}}{\partial X_{nij0}}$, represents the positive effect of land retirement on the ability of the ith parcel to trap sediment from upland parcels. The ith land parcel provides an external benefit to upland parcels by trapping a portion of their sediment and preventing it from being loaded into the water body.

The three terms that contribute to marginal value per acre suggest that two types of parcels make good candidates for land retirement. First, cropland parcels close to a water body, generating large amounts of eroded soil, and trapping sediment from upslope parcels make good candidates for land retirement. Second, upland parcels that would substantially improve the sediment trapping efficiencies of downslope parcels when taken out of crop production also make good candidates for a retirement program. In either case, land retirement is optimal if the forgone quasi-rent from crop production is low.

If the social planner were to set a uniform abatement standard (for example, the same percentage reduction in sediment loadings from baseline levels or the same absolute level of loadings) for all watersheds (for equity reasons or to reduce transactions costs) then the constraint in (5) must be replaced by N constraints as follows:

$$S_n^0 - \sum_{j=1}^{J_n}\sum_{i=1}^{I_{jn}}(1-\sum_{m=0}^{i-1} d_{n,i,i-m,j})\sum_{k=0}^{1} s_{nijk}X_{nijk} \geq \overline{A_n} \quad \text{for } n=1,\dots,N \qquad (7)$$

where A_n is the abatement standard for watershed n. The optimum solution now consists of N values of λ which differ because of differences in land characteristics among watersheds. Even though the same abatement goal can be achieved either with an aggregate abatement constraint or with a uniform standard for each watershed, having N abatement constraints will increase the costs of abatement according to the LeChatelier's principle.

The analysis above shows the fallacy of focusing only on retiring land in parcels with high on-site sediment generation irrespective of spatial location, that is, of retiring cropland parcels where $(s_{nij1}-s_{nij0})$ is large. By ignoring the other terms on the left-hand side of (6), which determine off-site sediment abatement and depend on the location of the parcel in the flow path, the abatement benefits provided by retiring an upslope parcel would be underestimated. This analysis also shows that treating deposition coefficients as fixed rather than endogenous implies that we are setting $\sum_{m=0}^{i-1}(\frac{\partial d_{n,i,i-m,j}}{\partial X_{nij0}})$ and $\sum_{m=1}^{I_j-i}\frac{\partial d_{n,i+m,i,j}}{\partial X_{nij0}}$ equal to zero and thus ignoring the second and third terms on the left-hand side of (6). This would result in an underestimate of the benefits of retiring a parcel. Furthermore by ignoring the effect of volume of run-off and land-use decisions of upslope parcels on deposition coefficients of downslope parcels we may fix the coefficients $d^*_{n,i,i-m,j}$ incorrectly and not obtain the correct estimate of the benefits of retiring a parcel. The empirical implementation of this framework in Khanna et al. (2003) and Yang et al. (2003) requires combining detailed data from a geographic information system on the characteristics and location of each land parcel in the watershed with a hydrological model that provides the sediment transport process for the watershed.

3. DESIGN OF GREEN PAYMENTS

3.1 Spatial heterogeneity and endogenous sediment deposition characteristics

The framework developed in Section 2.2 above can also be used to examine the design of rental payments to induce voluntary land retirement by land owners to achieve environmental goals at least cost. Equation (6) shows that a uniform dollar payment per ton of abatement (equal to the marginal value of abatement) across watersheds and across land parcels within a watershed can achieve cost-effectiveness. However, the rental payment per acre varies across land parcels. This per acre payment consists of two parts: the marginal value of abatement, which is the same for all parcels, and the contribution of the retired parcel to abatement per acre which depends on each parcel's capacity to reduce its own erosion, trap sediment from upslope parcels and improve the sediment trapping efficiencies of downslope parcels. Abatement per acre varies across parcels depending on their site-specific characteristics and the characteristics and land use decisions of all upslope and downslope parcels within a flow path.

A policy maker would need to know the relationships and parameters embedded in the model above to determine the parcel-specific rental payments per acre required to achieve the sediment abatement goal for the watershed. Khanna et al. (2003) find using numerical simulations that slope of the land parcel, distance from the water body and the soil quality of the parcel (which determines its productivity) are key observable characteristics that influence parcel specific rental payments. However, unobservable characteristics such as erodibility of the soil and the amount of sediment a parcel receives from upland parcels also influence a parcel's contribution to sediment abatement and its rental payments. Such a payment scheme might be difficult to implement and is likely to receive resistance from landowners because it is non-uniform and not transparent. A simpler alternative would be to set a rental cap for a watershed and enroll all land parcels offered by farmers at rental rates below the cap. The maximum payment or cap would equal the quasi-rent per acre of the marginal parcel that needs to be enrolled to achieve a given sediment abatement target. A cap creates incentives for landowners to retire parcels with low quasi-rents per acre, rather than parcels with high off-site abatement per dollar of quasi-rents foregone and is, therefore, unlikely to be cost-effective.

3.2 Uncertainty and irreversibility of technology adoption

The discussion above has focused on green payments to induce retirement of land from crop production. However, green payments to encourage farmers to adopt environmentally friendly practices are an equally important component of agro-environmental policy in the US. In the Farm Security and Rural Investment Act of 2002, EQIP has been funded at levels comparable to CRP[2]. Funding for EQIP almost doubles in 2003 to $700 million and then increases steadily through 2007 where it reaches $1.3 billion. These payments encourage adoption of improved nutrient, irrigation, manure, pest, and wildlife management practices, which may not otherwise be profitable. These payments are typically uniform payments per acre of land, structure, or facility and aimed at sharing 50 to 75 percent of the implementation costs to make it profitable for farmers to switch. Conservation practices may in some cases involve large fixed costs and irreversible investment decisions that must be made in the face of revenue uncertainty that can be characterized by a stochastic process. Additionally, farmers have the flexibility to decide not only whether or not to adopt but also when to adopt a conservation practice. For example, site-specific crop management (SSCM) provides an input-efficiency enhancing alternative to conventional methods by acquiring information about spatial variability in soil conditions and using it to target fertilizer applications to match that variability. SSCM relies on several inter-related components that include grid-based soil sampling and testing, yield monitors linked to satellite-based global positioning systems that provide geo-referenced information about the agronomic conditions and yields at various points in the field and variable rate technologies (VRT) that apply fertilizer at a varying rate across the field. SSCM has the potential to reduce over application of inputs such as nitrogen and nitrate run-off in at least some parts of the field. However, while the input cost savings and revenue increases occur in the future and the latter are uncertain due to uncertainty about prices, the fixed costs of adoption must be incurred at the time of adoption. Khanna et al. (2000) develop an option value framework to examine the adoption decision and its timing and to determine the green payment that induces immediate adoption of site-specific crop management (SSCM) instead of conventional farming methods. They also analyze how these payments need to vary across heterogeneous soil conditions and incorporate the value of waiting that arises due to the need to make an irreversible adoption decision under uncertainty.

They consider a price-taking profit-maximizing risk-neutral farmer operating a field of A acres. Soil fertility levels vary within the field and these differences are captured by dividing the field into $j=1,..J$ plots of size A_j acres. Suppose that the crop response function is $y_{jt}= f(s_{jt}, x_{jt})$ where y_{jt} is the per acre yield in the jth plot at time t which is a function of the soil fertility level per acre, s_{jt}, and applied input (fertilizer) per acre, x_{jt}. It is assumed that s_{jt} varies within the field with mean μ and variance σ_s^2. The soil fertility level could change over time due to carryover of unused fertilizer from one period to the next. It is also assumed that $f_s>0, f_x>0, f_{ss}<0, f_{xx}<0$. The farmer has a discrete choice between two technologies, Conventional and SSCM, denoted by superscripts C and S, where SSCM requires soil testing and adoption of VRT. Output price (P_t) is assumed to be changing over time and the farmer has expectations $E(P_t)$ of these prices in the future. Input price (w) is assumed to be fixed over time. The upfront cost of equipment, at $t=0$, required for SSCM is K. The lifetime of the equipment for SSCM is $\overline{T}$ years and discount rate is ρ. Under the conventional practices, the farmer lacks information about the distribution of soil fertility in the field but uses a small sample of soil tests to estimate the average soil fertility μ in the field. The farmer chooses the optimal uniform level of input-use per acre, x_t^C, for all J plots to maximize the discounted value of expected quasi-rents, π_0^C, taking the soil nutrient level to be at the average level μ in all plots:

$$\pi_0^C = \max_{x_t} \int_0^{\overline{T}} e^{-\rho t} A\big(E(P_t) f(x_t, \mu_t) - w x_t\big) dt \tag{8}$$

The profit-maximizing input rate is determined such that $\partial \pi_0^C / \partial x_t = E(P_t) f_x(x_t^C, \mu_t) - w = 0$.

Under SSCM the farmer invests in more intensive soil testing to learn about soil fertility levels in each of the J plots and applies the optimal (and spatially varying) input level in each plot. With SSCM, the farmer chooses the level of input application x_{jt}^S for each of the $j=1,..J$ plots knowing s_{jt} in each of the plots to maximize the discounted quasi-rents π_0^S where:

$$\pi_O^S = \sum_{j=1}^{J} \max_{x_{jt}} \int_0^{\overline{T}} e^{-\rho t} A_j \big(E(P_t) f(x_{jt}, s_{jt}) - w x_{jt}\big) dt \tag{9}$$

The input level at any point in the field is determined such that $\partial \pi_0^S / \partial x_{jt} = E(P_t) f_x(x_{jt}^S, s_{jt}) - w = 0$ and depends on the soil fertility level s at that point.

The impact of adoption on yield in the jth plot is approximated by a Taylor series expansion around the optimal level of input-use:

$$f(x_t^C, s_{jt}) - f(x_{jt}^S, s_{jt}) = f_x(x_t^C - x_{jt}^S) + f_{xx}(x_t^C - x_{jt}^S)^2 \tag{10}$$

The first term on the right hand side of (10) could be positive or negative depending on whether the plot has above average or below average fertility, while the last term on the right hand side of (10) is always negative since $f_{xx}<0$. This indicates that on plots with $x^S_{jt} > x^C_t$ yield is higher under SSCM than under conventional practices. On plots with $x^S_{jt} < x^C_t$ yields under SSCM are higher than under conventional practices if the second term on the right hand side in (10) is larger than the first term. The greater the variability in the soil fertility distribution in the field, the greater the magnitude of the second term and the greater the potential for yield gains with adoption even if the input application is reduced.

By substituting $f_x = w / E(P_t)$, rearranging terms and multiplying (10) by A_j one obtains the aggregate gains in expected quasi-rents with adoption of SSCM at a point in time t:

$$E(P_t)[Y_t^S - Y_t^C] - w(X_t^S - X_t^C) = -\sum_{j=1}^{J} E(P_t) f_{xx}(x_{jt}^C - x_{jt}^S)^2 > 0 \tag{11}$$

where X and Y represent aggregate levels of input-use and yield respectively. The term on the right hand side in (11) is always negative. As a result, over-application ($X^C_t > X^S_t$) under conventional methods relative to the optimal level leads to revenue gains that are lower than the increase in variable costs, while under-application ($X^C_t < X^S_t$) leads to revenue losses that are larger than the savings in variable input costs. The greater the variability in the fertility distribution, the greater the magnitude of the differential in (11). The higher the soil fertility level, the smaller is x^S_{jt} and the higher is f_{xx} and therefore the quasi-differential in (11). Hence, fields with higher soil fertility on average and greater variability in soil fertility are more likely to adopt. Thus gains due to adoption vary with the distribution of soil characteristics in the field and with the expected price of output. Assuming that applied nutrients and nutrients in the soil are perfect substitutes implies that $x_{jt}^S = x_t^C - (s_{jt} - \mu_t)$ if $s_{jt} - \mu_t \leq x_t^C$ and $x_{jt}^S = 0$ otherwise.

This together with (11) implies that the present value at $t=0$ of the differential in expected maximized quasi-rents π_o, can be written as:

$$\pi_o = \pi_o^{S*} - \pi_o^{C*} = \int_0^{\bar{T}} e^{-\rho t} [\sum_{j=1}^{J} E(P_t) f_{xx}(s_{jt} - \mu_t)^2] dt > 0 \tag{12}$$

where π_o^{S*} and π_o^{C*} represent the maximized NPV of quasi-rents under the two technologies. A farmer making the adoption decision based on the net present value (NPV) maximization criterion would compare π_o with the fixed costs of investment in SSCM (K). The gains in quasi-rents from adopting SSCM are always positive under certainty and risk neutrality and increase as the variability in soil fertility increases. Under the NPV rule, a farmer adopts SSCM at $t=0$ if $\pi_o \geq K$ or the rate of return is greater than ρ.

However, suppose that the farmer has the option of adopting at some instant $T=0,\ldots\hat{T}$ in the future where $\hat{T}$ is the planning horizon of the farmer. Let π_T denote the present value of the expected quasi-rent differential due to adoption at time T. Khanna et al. (2000) assume that π_T is stochastic and evolves according to a geometric Brownian motion with:

$$d\pi = \alpha\pi dt + \sigma\pi dz \tag{13}$$

where dz is the increment of a Wiener process; α is a proportional growth parameter and σ reflects the variance in the growth rate.

Let $\tilde{\pi}_T$ denote the threshold value of the discounted quasi-rent differential that is required for adoption to occur at time T. This value equals the incremental investment costs plus the value of the option to delay. Taking option values into account and assuming risk neutrality, they show that adoption would occur at T^* where:

$$\pi_{T^*} \geq \tilde{\pi}_{T^*} = \frac{\beta}{\beta - 1} K \text{ and } \beta = \frac{1}{2} - \frac{\alpha}{\sigma^2} + \sqrt{(\frac{\alpha}{\sigma^2} - \frac{1}{2})^2 + 2\frac{\rho}{\sigma^2}} \tag{14}$$

Thus the investment rule under uncertainty and irreversibility requires $\pi_T{}^*$ to be greater than K by a factor of $\beta/(\beta-1) > 1$, referred to as the option-value multiple for investment in SSCM. This multiple is a positive function of the growth rate, α, and the volatility of the growth rate σ and a negative function of the discount rate ρ. It varies with the characteristics of the soil distribution. In cases where it is not optimal for the farmer to adopt SSCM immediately under the option value approach, a cost-share subsidy could be used to accelerate adoption of SSCM to achieve greater pollution control. The cost-share subsidy C required for inducing immediate investment in SSCM when $\pi_o < \tilde{\pi}_o$ is determined such that: $C = K - \frac{\beta - 1}{\beta}\pi_o$ which is larger than the subsidy $K-\pi_o$ that would have been needed under the NPV criterion. Here π_0 is defined as above, and β is the option value multiple that can be shown to be greater than 1 (Dixit and Pindyck, 1994). It shows the extent to which the quasi-rent differential must exceed the investment costs before investment will occur, given uncertainty and sunk costs. Since π_o varies with the soil conditions on the field, the average level of soil fertility and the variability in the soil conditions in the field, the cost-share subsidy is also expected to vary across fields depending on their soil conditions.

A numerical simulation conducted by Khanna et al. (2000) shows that cost-share rates required to induce adoption on such soil distributions under the option-value approach can be 20% higher in some cases than those required to induce adoption under the NPV rule in a watershed in Illinois. Their analysis also shows how green payments need to vary across fields that differ in their average level of soil quality and in the variability of soil quality within the field. Isik et al. (2001) extend this analysis to show that when farmers have the possibility of adopting soil testing and VRT sequentially rather than as a package it is important to take that into account when determining the green payment required to induce adoption under uncertainty. The option value multiple for soil testing differs from that for VRT as does the subsidy level required to induce immediate investment in soil testing and VRT. Modeling SSCM as a package would underestimate the required subsidy for adoption of VRT while it would overestimate the required subsidy for soil testing.

A further extension of this framework by Isik (2004) shows the detrimental effect that uncertainty about the provision of a cost-share subsidy, in addition to revenue and cost uncertainty about the technology, has on the incentives and timing of adopting SSCM. An increase in the probability of a one-time subsidy that reduces the initial fixed costs of adoption creates incentives to delay immediate adoption while an increase in the probability that the subsidy, once in effect, could be withdrawn in the future accelerates adoption. Simulation results show that an effective approach to induce early adoption would be to offer a cost-share subsidy right away, threaten to remove it soon and commit to not restoring it again.

Kurkalova et al. (2002) empirically apply the option value framework to determine the adoption premium and the subsidy payments that would be needed to induce adoption in conservation tillage in Iowa. They find that even though the expected profit from conservation tillage is higher than that of conventional tillage, the difference is not large enough in all cases to cover the adoption premium due to the option value. Planners of green payment programs need to recognize the adoption premium needed to induce irreversible adoption decisions by reducing the sunk costs of adoption. Their analysis also shows that green payments need to vary with the heterogeneous characteristics of farmers, which influence the profitability of conservation tillage and the adoption premium.

3.3 Uncertainty about technology and risk aversion

Using a one-period model, Isik and Khanna (2002; 2003) extend the approach presented above to examine the impact of risk aversion on adoption decisions for SSCM when errors exist in the measurement of soil nutrient levels. Under SSCM the farmer considers the production function to be stochastic and represented by: $y_j = f(x_j, s_j + s_j\varepsilon)$ where ε is a random variable with mean zero and variance σ_ε^2. This leads to uncertainty about the returns from adoption of SSCM relative to conventional methods that are simply based on the average soil fertility level, assumed to be known with certainty[3]. The utility-maximizing adoption decision is obtained through a two-stage decision process. The farmer first obtains the utility-maximizing level of input use with each technology and then compares the maximized expected utility with adoption of VRT and with the conventional practices. The farmer adopts SSCM if the expected utility from its adoption is greater than that of the conventional practices.

Isik and Khanna (2002; 2003) show that uncertainty and risk aversion can affect input use as well as the adoption decision. The impact on input use arises from the existence of a marginal risk premium which creates a wedge between the input cost and the expected marginal product at the optimal level of input use. The marginal risk premium could be positive (negative) and the input is risk increasing (reducing) if f_{xs} > (<) 0. A risk-averse farmer uses more of an input having a negative marginal risk premium. Using an exponential utility function, the authors obtain the cost-share subsidy CS that should be offered to farmers to induce adoption:

$$CS = K + \frac{\phi v^2}{2} + \frac{AP}{2f_{xx}}\left[\sigma_s^2 (f_{xz})^2 - \frac{(f_x)^2}{A}\sum_{j\in J} A_j \left(\frac{(R/Pf_{xj})^2}{(1-R/Pf_{xj})^2}\right)\right] \qquad (15)$$

where $\frac{\phi}{2}v^2$ is the risk premium with ϕ representing the degree of risk aversion and v^2 the variability in the returns from adopting SSCM. The last term on the right hand side in (15) is the quasi-rent differential which is positive if SSCM leads to lower quasi-rents. Under uncertainty and risk aversion, SSCM may not always lead to an increase in the quasi-rents. Adoption could lead to a reduction in the quasi-rent differential if the impact of risk aversion and uncertainty on input use is very high and the variability of soil nutrients σ_s^2 is very low. The cost-share subsidy is the sum of this loss in quasi-rents due to SSCM, the costs of investment in SSCM and the risk premium.

Ignoring soil nutrient uncertainty and risk aversion, a planner would underestimate the required cost-share subsidy to induce adoption. The cost-share required for inducing adoption of SSCM under soil nutrient uncertainty and risk aversion increases with an increase in cost of adoption, risk aversion and variability of returns, and it decreases with an increase in variability of soil conditions. Cost-share subsidies to induce adoption and reduce pollution are effective if uncertainty is low, the degree of risk aversion is low, and fertilizer is a risk increasing input. These subsidies need to be larger for smaller farms and those with low spatial variability in soil nutrient distribution. While a cost-share subsidy that includes a risk premium can induce adoption of SSCM, it would not influence the amount of pollution generated after adoption under soil nutrient uncertainty and risk aversion. To achieve a reduction in nitrogen pollution through adoption of VRT in the presence of uncertainty it may be necessary to supplement cost-share subsidies with insurance policies that reduce the risks to profits and yields and reduce the incentives for farmers to over-apply risk-decreasing inputs. This analysis shows the importance of understanding the impact of risk, uncertainty and spatial variability on quasi-rents and input-use with adoption for the design of policies that are effective in reducing non-point pollution by inducing adoption of SSCM.

Parks (1995) examined the impact of uncertainty about land use benefits on the incentives of a risk-averse landowner with a given stock of land to convert marginal agricultural lands into forests. He showed that annualized subsidies to convert to forests must outweigh the sum of both risk and capital gains (due to changes in the relative values of forest land to agricultural land in the future) besides the difference in annual returns. Furthermore, technical assistance, which reduces a landowner's uncertainty about forestland use benefits, increases the conversion of marginal agricultural land to forested land.

4. EFFICIENCY OF GREEN PAYMENT POLICIES

A tax on the pollutant to be abated would be a cost-effective approach to achieve aggregate environmental standards in the presence of heterogeneity among polluting sources when there is perfect information about costs of abatement and pollution can be measured (Griffin and Bromley, 1982). The implementation costs associated with mandatory policies do not arise if green payment programs are used to induce voluntary adoption of (otherwise unprofitable) conservation practices that reduce environmental damage. Farmers are heterogeneous and suffer varying levels of loss of profits from adoption of the conservation practice. The government can provide green payments to cover the losses from adoption and provide services such as information and technical assistance that reduce these losses. The cost to the government of providing these services depends on the degree of non-rivalness of the service. The provision of green payments imposes deadweight losses associated with using costly tax revenues.

Wu and Babcock (1999) compare the efficiency of a mandatory policy where a fine is large enough to ensure that all farmers adopt a conservation practice with that of a uniformly applied green payment per acre set equal to the largest profit loss per acre among farmers to ensure participation by all farmers. They show that a green payment program is more efficient than a mandatory tax policy if

$$\lambda E^{v} - \Delta C < \Delta R \tag{16}$$

where λE^{v} is the deadweight loss of government expenditure on the voluntary program, ΔC is the extent to which the green payment program lowers the costs of adoption of the conservation practice and ΔR is the difference in enforcement costs between the mandatory and green payment programs. If the deadweight loss of the subsidy is zero, a voluntary program gains the comparative advantage because of the enforcement costs of a mandatory program.

Wu and Babcock (1999) also show that as more small farms are targeted for conservation practices, the comparative advantage of a green payment program increases because duplicated private effort is avoided by government technical and information services that lower costs of abatement. As government services increase and costs of adoption decrease, the need for subsidies decreases, thus reducing the deadweight loss of direct payments. However, as program acreage increases, the relative efficiency of green payment programs depends on how fast monitoring and enforcement costs increase under the two approaches. Finally, the authors show that the relative efficiency of a voluntary program increases when the degree of rivalness of government services decreases, government services cost less than equivalent private services and/or enforcement costs of mandatory programs increase. However, they assume no exit or entry of land; a key source of divergence between a tax and a subsidy policy. The existence of heterogeneous land quality implies that while a tax might induce exit, a subsidy could induce entry by idle marginal land currently not in production. These slippage effects can be substantial as shown by Wu (2000) in the case of the Conservation Reserve Program. Additionally, Wu and Babcock (1999) assume that both the mandatory program and the green payment program induce all farmers to adopt the conservation practice. Since the costs of adoption and the environmental benefits from adoption of conservation methods is likely to vary across heterogeneous farmers, it may not be socially desirable to induce universal adoption with a green payment program.

Khanna et al. (2002) examine the cost-effectiveness of alternative green payment policies relative to a pollution tax to achieve the same level of pollution control. They extend the framework developed by Caswell and Zilberman (1986) and utilized by Hellegers (this book) in which pollution generated by a land parcel can decrease in three ways - switching to a conservation technology that increases input efficiency and reduces pollution intensity (switching effect), reducing polluting input use with a given technology (intensive margin effect) and exiting from the industry (extensive margin effect). The cost-effective combination of these three effects depends on the strength of the input-saving and pollution reducing characteristics of the conservation technology, the responsiveness of input use to prices, and the fixed costs of adopting a conservation technology. The green payment policies considered in their paper are cost-share subsidies that partially offset the fixed costs of adopting a conservation technology, input reduction subsidies that reduce the use of a polluting input, and combinations of the two types of subsidy payments. Two versions of each policy are examined; one where entitlement is restricted to currently operating units and the other that allows unrestricted entry.

These green payment policies differ in the incentives they provide for alternative ways of pollution control and therefore diverge in their costs of abatement and production response, while achieving the same level of pollution control. A cost-share subsidy with unrestricted entry achieves a reduction in pollution simply through the switching effect and does not affect the intensive margin. An input-reduction subsidy achieves a reduction in pollution through both a switching effect and an intensive margin effect. While the tax induces microunits to exit the industry, an input reduction subsidy and/or the cost-share subsidy induce entry. This tends to increase the pollution generated (relative to a restricted input reduction subsidy and/or the cost-share subsidy) and necessitates larger subsidy rates than the restricted versions of these policies to achieve the targeted level of abatement. Additionally, unlike a uniform input-reduction subsidy that raises input price uniformly for all land parcels, a pollution tax raises input price more for land that is of lower quality and for land under the traditional technology; hence the pollution tax achieves greater targeting and cost-effectiveness.

Since a cost-share subsidy has no intensive margin effect and achieves pollution control only through technology switching it is an effective policy tool for reducing pollution only if the technology switching effect is large and if the conservation technology has a large pollution reducing effect. However, this could be a very costly strategy for abatement if the conservation technology has high capital costs and input use is responsive to a tax, making input use reduction a preferable method for pollution control. Although, in their framework, green payment policies are second best to a pollution tax, the inefficiency of a restricted cost-share policy may not be too large if the intensive and extensive margin effects of a pollution tax are small because tax payments are a small share of total revenue. Additionally, if the extent of heterogeneity among microunits is small, then a uniform input reduction subsidy could achieve intensive and switching effects that are very close to those of a pollution tax. A combined policy with restricted cost-share and input reduction subsidy is the closest in replicating the incentives provided by a pollution tax policy. A numerical simulation to analyze the implications of alternative policies for reducing polluted drainage from irrigated cotton production in the San Joaquin Valley in California shows that the difference in the costs of abatement between the least-cost tax policy and a restricted combined green payment policy is only 1.2% of the base level of social welfare in the study region. The unrestricted version of this combined policy costs 10% of the base level of social welfare more than the least cost policy.

5. CONCLUSIONS

There is growing reliance on subsidy/green payments to agricultural landowners who reduce nonpoint-source pollution and enhance environmental services either by retiring land from crop production and converting it to forests or permanent cover or by adopting environmentally friendly technologies and less polluting input management strategies. This has led to a number of studies examining the implications of alternative targeting mechanisms for these payments, their design and their effectiveness under conditions of spatial, temporal, and farmer heterogeneity and uncertainty about prices, weather and performance of the technologies whose adoption these payments seek to induce.

Alternative decision rules for selecting land to be retired from crop production can result in considerably different levels of environmental benefits depending on the distribution of environmental benefits and opportunity costs of land retirement in a region and the correlation between the two. Implementing a rule that maximizes the benefit to cost ratio from land retirement can be highly information intensive and require site-specific information about land characteristics. In the case of run-off related water quality problems, it also requires linking on-site pollution generation with off-site pollution loadings. For some pollution problems such as those caused by sediment run-off, this linkage cannot be specified exogenously but depends on the characteristics and land use decisions by neighboring parcels. In the absence of ways to directly observe or measure each land parcels contribution to run-off and off-site pollution loadings, hydrological models together with GIS can be used as substitutes to specify pollution generation and transport functions and model the interdependencies among land parcel choices and consequences. These can be combined with a landowner's decision model to examine the land that should be targeted to receive green payments to switch to environmentally friendly practices to achieve environmental benefits cost effectively. The implementation of a green payment policy then requires an additional step, namely, designing a payment policy that would replicate a social planner's choices in a decentralized setting where landowners make voluntary decisions to adopt environmentally friendly practices in response to market based incentives. With uniform mixing of pollutants in the program area, green payments per ton of pollutant abated need to be uniform across watersheds in that area in order to achieve abatement targets cost-effectively. Rental payments per acre of land to induce retirement of land from crop production and cost-share rates per acre of land to induce adoption of environmentally friendly technologies need to, however, vary spatially with the heterogeneity in land characteristics. The use of these simulation models to design green payment policies ignores the stochastic factors that influence nonpoint pollution and assumes perfect information about landowners' choices about management practices. However, they provide a plausible approach useful for practical implementation not only of pollution-based green payment policies but even for other incentive based policies for nonpoint pollution control such as tradable permits and taxes.

In practice, regulators may lack the information or funds to develop suitable simulation models to link on-site pollution generation to expected off-site loadings and instead use green payments to induce producers to change observable behavior by adopting environmentally friendly technologies. These technologies may require farmers to incur high fixed costs at the time of adoption and to make an irreversible investment decision. The benefits of adoption may however be uncertain, either due to uncertainty about crop prices or the weather or about the regulatory/green payment regime in the future. Additionally, farmers have flexibility in choosing not only whether or not to adopt but when to adopt the technology. In such cases, the level of green payments required to induce immediate adoption of such technologies would need to compensate risk neutral farmers not only for the loss in discounted profits from adoption but also for the lost option value of waiting before making an irreversible decision. Furthermore, if farmers have a choice of whether to adopt the technology as a complete package or to adopt its components sequentially, green payments may need to be offered selectively on some components only.

Weather uncertainty and/or uncertainty about the performance of these technologies may also affect the private benefits of environmentally friendly technologies relative to conventional technologies. This coupled with risk averse behavior can diminish incentives to adopt such technologies. Green payment policies then need to incorporate a risk premium to induce adoption. However, in the presence of risk aversion and uncertainty, adoption may not lead to the desired reduction in use of polluting inputs. Green payments may need to be supplemented by insurance programs that cover the risks to profits and yields and reduce incentives to over-apply polluting inputs.

To be effective, regulators not only need to determine the level of green payments based on the contribution of a land parcel to pollution abatement, the land owner's option values and risk premium but also how to vary them in response to heterogeneity of soil conditions, location, topography and heterogeneity in risk attitudes, discount rates and expectations about prices, costs and regulations in the future. Uniform green payment policies, such as a technology cost-share or an input reduction subsidy are likely to be less efficient than an emissions tax policy if the latter could be implemented. This is because the latter achieves pollution reduction through three mechanisms, a negative extensive margin effect, an intensive margin effect and a technology switching effect while a green payment policy may rely at most on the latter two effects and have a positive (instead of negative) extensive margin effect. By restricting green payments to lands that are already under production and by combining both cost-share subsides for technology adoption and input-reduction subsidies, it is possible to limit efficiency losses associated with green payment programs. Additionally, if emission tax policies have enforcement costs associated with them and if the deadweight losses of raising funds to provide green payments are low then green payment policies may even become more efficient than emission tax policies. In summary, the effectiveness of green payment programs for environmental protection depends on the social cost of funds, on the regulator's ability to target payments appropriately, to prevent slippage, to choose payment levels appropriately and to vary them in response to heterogeneity in soil conditions, farmer preferences and environmental benefits.

Notes

1. There are also several studies examining the optimal design of green payment policies when regulators have asymmetric information about the types of farmers (Wu and Babcock, 1995; Smith, 1995). In that case farmers may have an incentive to misrepresent their type to obtain favorable combination of production practices and green payments and these studies design green payment policies that provide incentives for farmers to self-select payments and practices intended for their type. In this chapter we only focus on issues related to green payments when regulators have complete information.

2. Economic Research Service. 2003 "The 2002 Farm Bill: Provisions and Economic Implications."http://www.ers.usda.gov/Features/FarmBill/Titles/TitleIIConservation.htm. Accessed June 6, 2003.

3. Isik and Khanna (2002) assume that only returns from SSCM are stochastic due to uncertainty about soil nutrient levels. This assumption is relaxed in Isik and Khanna (2003) to allow for uncertainty about soil nutrient levels affecting returns from conventional methods as well.

References

Babcock B.A., P.G. Lakshminarayan, J. Wu, and D. Zilberman, 1996. 'The Economics of a Public Fund for Environmental Amenities: A

Study of CRP Contracts', *American Journal of Agricultural Economics*, 78(4), 961-971.

Babcock B.A., P.G. Lakshminarayan, J. Wu, and D. Zilberman, 1997. 'Targeting Tools for the Purchase of Environmental Amenities', *Land Economics*, 73(3), 325-339.

Braden, J.B., G.V. Johnson, A. Bouzaher, and D. Miltz, 1989. 'Optimal Spatial Management of Agricultural Pollution', *American Journal of Agricultural Economics,* 61(5), 404-413.

Carpentier, C.L., D.J. Bosch, and S.S. Batie, 1998. 'Using Spatial Information to Reduce Costs of Controlling Agricultural Nonpoint Source Pollution', *Agricultural and Resources Review*. 27(1), 72-84.

Caswell, M.F. and D. Zilberman, 1986. 'The Effects of Well Depth And Land Quality On The Choice Of Irrigation Technology,' *American Journal of Agricultural Economics,* 68(4), 798-811.

Dixit, A. K. and R.S. Pindyck, 1994. *Investment Under Uncertainty,* Princeton University Press, Princeton, NJ.

Griffin, R.C. and D.W. Bromley, 1982. 'Agricultural Runoff as Nonpoint Externality: A Theoretical Development', *American Journal of Agricultural Economics* 64(3), 547-552.

Hellegers, P. J.G.J, 'The Impact of Recovery of Irrigation Water Losses on the Choice of Irrigation Technology with Heterogeneous Land Quality and Different Crops' in *Frontiers in Water Economics* edited by Renan Goetz and Dolors Berga (this book).

Horan, R.D and J.S. Shortle, 2001. 'The Economics of Nonpoint Pollution Control', *Journal of Economic Surveys*, 15(3), 255-289.

Isik, M., 2004. 'Incentives for Technology Adoption Under Environmental Policy Uncertainty: Implications for Green Payment Programs', *Environmental and Resource Economics,* 27(3), 247-263.

Isik, M. and M. Khanna, 2002. 'Uncertainty and Spatial Variability: Incentives for Variable Rate Technology Adoption in Agriculture', *Risk, Decision and Policy*, 7(3), 249-265.

Isik, M. and M. Khanna, 2003. 'Stochastic Technology, Risk Preferences and Adoption of Site-specific Technologies', *American Journal of Agricultural Economics*, 85(2), 305-317.

Isik, M., M. Khanna, and A. Winter-Nelson, 2001. 'Sequential Investment in Site-Specific Crop Management Under Output Price Uncertainty', *Journal of Agricultural and Resource Economics* 26(1), 212-229.

Khanna, M., M. Isik, and A. Winter-Nelson, 2000. 'Investment in Site-Specific Crop Management under Uncertainty: Implications for Nitrate Pollution Control and Environmental Policy', *Agricultural Economics*, 24(1), 9-21.

Khanna, M., M. Isik, and D. Zilberman, 2002. 'Cost-Effectiveness of Alternative Green Payment Policies for Conservation Technology Adoption with Heterogeneous Land Quality', *Agricultural Economics*, 21(2), 157-174.

Khanna, M., W. Yang, R. Farnsworth, and H. Onal, 2003. 'Optimal Targeting of CREP to Improve Water Quality: Determining Land Rental Offers with Endogenous Sediment Deposition Coefficients', *American Journal of Agricultural Economics*, 85(3), 538-553.

Kurkalova, L, C. Kling, and J. Zhao, 2002. 'The Subsidy for Adopting Conservation Tillage: Estimation from Observed Behavior', working paper, Iowa State University.

Lintner, A. M. and A. Weersink, 1999. 'Endogenous Transport Coefficients: Implications for Improving Water Quality from Multi-contaminants in an Agricultural Watershed', *Environmental and Resource Economics*, 14(2), 269-296.

Parks, P., 1995. 'Explaining 'Irrational' Land Use: Risk Aversion and Marginal Agricultural Land', *Journal of Environmental Economics and Management*, 28, 34-47.

Prato, T. and S. Wu, 1996. 'Alternative Spatial Criteria for Targeting Soil and Water Quality Improvements in an Agricultural Watershed', *Review of Agricultural Economics*, 18, 293-301.

Ribaudo, M.O., 1989. 'Targeting the Conservation Reserve Program to Maximize Water Quality Benefits', *Land Economics*, 65(4), 320-332.

Smith, R.B.W., 1995. 'The Conservation Reserve Program as a Least-Cost Land Retirement Mechanism', *American Journal of Agricultural Economics*, 77(1), 93-105.

USEPA, 2001. 'The United States Experience with Economic Incentives for Protecting the Environment.' EPA-240-R-01-001. Washington, D.C.

Wu, J., 2000. 'Slippage Effects of the Conservation Reserve Program, *American Journal of Agricultural Economics*', 82(4), 979-992.

Wu, J. and B.A. Babcock, 1999. 'The Relative Efficiency of Voluntary vs. Mandatory Environmental Regulations', *Journal of Environmental Economics and Management*, 38, 158-175.

Wu, J. and B.A. Babcock, 1995. 'Optimal Design of a Voluntary Green Payment Program under Asymmetric Information', *Journal of Agricultural and Resource Economics*, 20, 316-327.

Yang, W., M. Khanna, R. Farnsworth, and H. Onal, 2003. 'Integrating Economic, Environmental and GIS Modeling to Target Cost Effective Land Retirement in Multiple Watersheds', *Ecological Economics*, 46, 249-267.

Chapter 11

CONJUNCTIVE USE OF SURFACE AND GROUNDWATER WITH QUALITY CONSIDERATIONS

Catarina Roseta-Palma
ISCTE, Portugal

1. INTRODUCTION

In most water systems groundwater is not used on its own but rather as a complement of whatever surface water supplies are available (rainfall, stream flows, surface water reservoirs). Accordingly, the literature that analyses management of groundwater stocks has included conjunctive use from the beginning (see the seminal article by Burt, 1967; and the reviews on the topic by Provencher, 1995; and Tsur, 1997). With a set of simple assumptions, such as that surface water is constant, cheaper, and that surface and groundwater are perfect substitutes, deterministic conjunctive use models are not much more complicated than groundwater-only management models. The main difference is that groundwater is used only after the given endowment of surface water has been exhausted.

A natural extension that brings these models closer to reality is to consider stochastic surface water supplies, highlighting the role played by groundwater in protecting users against uncertainty. Tsur (1990) studies the buffer role of groundwater in a static setting and shows that it is positive under standard concavity assumptions of the benefit function, so that groundwater is more highly valued when surface water varies than when it is constant. Tsur and Graham-Tomasi (1991) provide a similar analysis for a dynamic setting, although in this case the proof of positive buffer values requires more restrictive assumptions (namely, that marginal benefits are convex). Knapp and Olson (1995) also consider surface water variability; their paper analyses decision rules and establishes conditions for convergence of extraction and stock to limiting probability distributions using lattice programming, which is a useful method in problems where the value function associated with the dynamic programming problem is not concave. Provencher and Burt (1994) consider a two period model of surface water variability with risk averse firms to identify the risk externality associated with common property situations.

However, there is one aspect of water use that seems to have been somewhat overlooked in typical conjunctive use models. Considering that water quality is a relevant parameter in many regions, the fact that there are two alternative sources of water gains a new significance, since there is no guarantee that both sources will be of comparable quality. Therefore, the benefit for users of using one unit of surface water may not be the same as that of using one unit of groundwater. Yet they should still be regarded as part of one single management system, except in the few extreme situations where only one source of water is explored.

There are some examples of deterministic models of joint quantity-quality management of groundwater in economic literature, but the only case where a conjunctive system is considered is the salinity model of Dinar (1994) and Dinar and Xepapadeas (1998). There is also a paper on drainage problems by Tsur (1991) which briefly touches the issue. The Tsur and Zemel (1997) paper on irreversibility has a stochastic element (the size of stock below which groundwater use becomes unfeasible is unknown), but it does not consider conjunctive use. Two other papers that consider uncertainty but not conjunctive use are Fisher and Rubio (1997), where the recharge flow is variable and the maximum size of water stock depends on how much capital is invested, and Rubio and Castro (1996), where there is both recharge and demand uncertainty.

This article analyses the implications of considering a conjunctive ground and surface water system where water quality varies according to source, with and without uncertainty in hydrological parameters. A simple, static model of conjunctive use is introduced in section 2 to illustrate the issues that ensue from the inclusion of a quality parameter in the water revenue function. Results are compared to those of Tsur (1990). Finally, a dynamic model of groundwater evolution is presented, both in the standard deterministic version and in a stochastic version that uses methods similar to those in Fisher and Rubio (1997) or Rubio (1992).

2. STATIC MODEL

Users of the water (such as farmers) are assumed to maximize their profit by choosing the amount of water they want to apply. Surface water is exogenous, so that by choosing total water use the amount of groundwater to be pumped is established. There is a fixed unit cost of pumping, z, surface water, s, is provided at no cost, and there is a water revenue function $y(w, C)$ which depends on total water used, w, and on the concentration of some undesirable pollutant in that water, C. The maximum profit, Π, is:

$$\Pi = \max_{\{w\}} \quad y(w, C) - z(w - s) \tag{1}$$

Under the usual assumptions on the revenue function (namely, considering that the derivatives of y have the following properties: $y_w > 0, y_{ww} < 0, y_C < 0, y_{wC} \leq 0$)[1] this looks like a simple conjunctive use problem. However, even if the pollutant concentration levels in surface water and groundwater are both exogenous, C will be a weighted average of the two, thus it will be endogenous:

$$\begin{aligned} C &= C^s \frac{s}{w} + C^g \frac{w - s}{w} \\ &= C^g - \frac{s(C^g - C^s)}{w} \end{aligned} \tag{2}$$

where C^s and C^g are respectively pollutant concentrations in surface and groundwater.[2]

This introduces two different sorts of new problems: first, the likelihood of getting a differentiable and concave objective function using only the intuitive assumptions presented above is a lot smaller, so that second order conditions will be more difficult to check. All cases that will be analyzed in this chapter assume that the functional objective is well behaved: concavity is satisfied, and $w > s$ (ensuring differentiability in the relevant range of w). Situations where excess water is harmful, such as floods, although possible, are ruled out. It is considered that the amount of surface water is never too large, so that the last unit of water received is still revenue-increasing.

Second, the optimal choice of water will vary with the amount of available surface water, which is something that did not happen in quantity-only static conjunctive models. To show this, consider the first order condition for problem (1):

$$\begin{aligned} & y_w + y_C C_w = z \\ \iff \quad & y_w + y_C \tfrac{s(C^g - C^s)}{w^2} = z \end{aligned} \tag{3}$$

Thus the marginal benefit of pumping has two terms: the first one is the direct impact on production of pumping additional water, and the second is the impact on production through the effect on water quality. Note that this term is positive if groundwater is less contaminated than surface water and negative otherwise.

Equation (3) implicitly defines the optimal water decision, $w^*(s, C^g, C^s, z)$, so that:

$$w_s^* = -\frac{y_{wC}\frac{-(C^g - C^s)}{w} + y_{CC}\frac{-s(C^g - C^s)^2}{w^3} + y_C\frac{(C^g - C^s)}{w^2}}{y_{ww} + 2y_{wC}C_w + y_{CC}\left(C_w\right)^2 + y_C C_{ww}} \tag{4}$$

Note that if $C^g = C^s = C, w_s^* = 0$ and the traditional conjunctive use model holds. In that case, if surface water fluctuates, groundwater is simply pumped so as to keep total water used constant (ie. stabilizing water consumption).[3] If $C^g \neq C^s$, however, optimal water consumption is not stable when s varies. It may increase or decrease, depending on the sign of the numerator in equation (4) (the denominator is negative by the assumption of concavity).

EXAMPLE 1 *Suppose s is rainfall; then it should be true that $C^g - C^s > 0$. If $y_{CC} = 0$ and $y_{wC} = 0$, then $w_s^* < 0$. Thus, an increase in rainfall will decrease total water used. The reason is that an increase in s increases the negative impact on concentration of additional pumped units of water (C_w increases), so that if w remained constant the marginal benefit of pumping an extra unit would be lower than the marginal cost. This requires optimal w to decrease. If $y_{CC} > 0$ (ie. benefit is convex in C) the effect of a larger s would be even stronger, and w would decrease even more, whereas if $y_{CC} < 0$ then w would decrease less or even increase. If $y_{wC} < 0$, on the other hand, there is an extra increase in the marginal benefit of using water due to the higher availability of the better quality water (s), so that w_s^* tends to increase, although it may be positive or negative.*

Performing comparative static analysis with the remaining parameters of the model highlights some other interesting properties of the optimal water choice. Denoting $\zeta = y_{ww} + 2y_{wC}C_w + y_{CC}(C_w)^2 + y_C C_{ww}$, and recalling that $\zeta < 0$, the following results are obtained:

$$w_z^* = \frac{1}{\zeta} < 0 \tag{5}$$

$$w_{C^g}^* = -\frac{y_{wC}\frac{(w-s)}{w} + y_{CC}\frac{(w-s)}{w}\frac{s(C^g - C^s)}{w^2} + y_C\frac{s}{w^2}}{\zeta} \tag{6}$$

$$w_{C^s}^* = -\frac{y_{wC}\frac{s}{w} + y_{CC}\frac{s}{w}\frac{s(C^g - C^s)}{w^2} - y_C\frac{s}{w^2}}{\zeta} \tag{7}$$

Note that an increase (decrease) in w^* corresponds to an increase (decrease) in pumped water, since surface water is now being held constant. Thus, equation (5) shows that, as expected, less water is pumped when pumping costs increase. However, equations (6) and (7) are ambiguous. The optimal reaction to a higher level of contamination in either type of water is undetermined, depending again on y_{CC}, y_{wC} and $(C^g - C^s)$. If the second order derivatives are zero, then $w_{C^g}^* < 0$ (less groundwater is pumped when its quality deteriorates[4]) and $w_{C^s}^* > 0$ (more groundwater is pumped to compensate a fall in surface water quality). These results seem reasonable, but they do not hold in general. For instance, it is actually possible for more groundwater to be pumped even though its quality has fallen; notice that the numerator of equation (6) can be written as $\left(\frac{d(y_C)}{dw}\right)\frac{w-s}{w} + y_C\frac{s}{w^2}$. The second term is negative, so that $w_{C^g}^* > 0$ requires

$\frac{d(y_C)}{dw} > 0$, which implies that the (negative) impact of quality on revenue will increase (ie. become less negative), increasing the attractiveness of pumping extra water. If this effect is strong enough, more water will be pumped.

2.1 Surface water variability

In many geographical regions surface water supplies fluctuate greatly between periods. In Portugal, for example, the available water supplies in a dry year can be as little as one third of their average values (see INAG, 2001). It has already been remarked in the previous section that the optimal consumption of water in a simple model, without quality considerations, is invariant with surface water. Accordingly, in the presence of a stochastic surface water supply, groundwater will be used to complement surface water so that total water use remains constant. When there are quality differences between the two types of water, this result no longer holds, and groundwater use may fluctuate more or less than surface water. Moreover, the impact on groundwater use will depend on whether pumping decisions are made before (ex ante) or after (ex post) the exact realization of surface water is known. The latter is the more realistic assumption for most systems, thus it will be the one pursued here.

Surface water variability is introduced into a static conjunctive use model in Tsur (1990). As he notes, in a static model where decisions are made ex post \the uncertainty of water supplies is really an instability". He compares the value of groundwater when s is a random variable to its value when s is fixed at the mean ($\overline{s}$), naming the difference the buffer value of groundwater. He shows that the buffer value is positive as long as the water benefit function is concave. In this section the same concept is applied to the case where there are quality differences. To ensure differentiability for any s, it must be assumed that desired water use will be greater than the highest admissible value for s.

By definition, the buffer value of groundwater is given by:

$$\begin{aligned} BV &= E\left\{y(w^*(s), C(w^*(s), s)) - y(s, C^s) - z(w^*(s) - s)\right\} \\ &\quad - \left[y(w^*(\overline{s}), C(w^*(\overline{s}), \overline{s})) - y(\overline{s}, C^s) - z(w^*(\overline{s}) - \overline{s})\right] \\ &= \underbrace{y(\overline{s}, C^s) - E\left\{y(s, C^s)\right\}}_{1} \; \underbrace{+E\left\{\Pi(s)\right\} - \Pi(\overline{s})}_{2} \end{aligned} \tag{8}$$

By Jensen's inequality, the first term is positive under simple concavity of y in w. In fact, in Tsur's model the buffer value is exactly equal to this term,[5] since other terms are zero when w^* is independent of surface water. Thus he concludes that the buffer value is always positive. In our case, to ascertain the sign of the buffer value the curvature properties of Π have to be investigated.

Using the envelope theorem and recalling expressions (1) and (2):

$$\Pi_s = y_C \frac{-(C^g - C^s)}{w} + z \tag{9}$$

The sign of Π_s depends on which source of water is more contaminated, with $\Pi_s > 0$ whenever surface water is the relatively cleaner source (ie. $(C^g - C^s) > 0$), and $\Pi_s < 0$ when surface water is the relatively more polluted source. It should be stressed that the increase in maximum profit depends only on the relative contamination of surface water, not on its absolute value.[6] As for second order conditions, differentiating (9) yields:

$$\Pi_{ss} = \left(y_{CC} \frac{-(C^g - C^s)}{w} + y_{wC} w_s^* \right) \frac{-(C^g - C^s)}{w} + y_C \frac{(C^g - C^s)}{w^2} w_s^* \tag{10}$$

If $\Pi_{ss} > 0$, then the buffer value is always positive and it is greater than in the no quality model. Otherwise its sign is undetermined. Although the sign of Π_{ss} cannot always be ascertained, it is possible to check that it is positive for the case of $y_{CC} = 0$. In this case, taking into account that w_s^* is given by equation (4), expression (10) can then be rewritten as:

$$\Pi_{ss} = -\frac{1}{\zeta} \left[\left(\frac{y_C}{w} - y_{wC} \right) \frac{(C^g - C^s)}{w} \right]^2 > 0$$

On the other hand, if $\Pi_{ss} < 0$, then the buffer value is lower than in the no quality case, and it cannot be guaranteed that its sign will be positive. Thus the incorporation of quality differences raises new questions on the buffer role of groundwater.

3. DYNAMIC WATER STOCK EVOLUTION

3.1 Optimal choices under certainty

Considering that the groundwater stock is not constant implies that pumping cost is not constant either. Moreover, when taking aquifer dynamics into account all users of the aquifer system must be considered simultaneously. It is assumed that there are M identical agents exploiting a single stock of groundwater, which contains $G(t)$ units of recoverable water and is characterised by a flat bottom and perpendicular sides. The aquifer receives a constant recharge, R. The unit cost of groundwater extraction, denoted by $z(G(t))$, depends negatively on the size of the groundwater stock and the cost increase per unit depleted is higher the lower the remaining stock (i.e. $z(G(t))$ is decreasing and convex). A percentage α of the applied water returns to the aquifer, so that $G(t)$ evolves according to:

$$\dot{G} = M \left[-(w(t) - s) + \alpha w(t) \right] + R \tag{11}$$

Depending on the source of surface water being considered, it would also be possible for its amount and quality to be stock variables (surface water reservoirs, lakes). However, that would bring additional complexity to the model without bringing new insights, so in this chapter surface water is always considered a flow variable in the sense that it is used up immediately.[7] s, C^s and C^g are known, constant values.

Optimal use of the aquifer requires (let $M = 1$ for the moment since it does not affect optimal choices):

$$\max_{\{w(t)\}} \int_0^\infty \left[y(w(t), C(t)) - z(G(t))(w(t) - s)\right] e^{-\rho t} dt \qquad (12)$$

subject to equation (11) and to non-negativity restrictions on $w(t)$ and $G(t)$, as well as an initial condition $G(0) = G_0$. The current value Hamiltonian for this problem is:[8]

$$H = \left[y(w, C^g - \frac{s(C^g - C^s)}{w}) - z(G)(w - s)\right] + \lambda\left(\dot{G}\right)$$

Letting π denote instantaneous profit, $\pi_w = y_w + y_C \frac{s(C^g - C^s)}{w^2} - z(G)$ and first order conditions for interior solutions can be stated as:

$$\pi_w = \lambda(1 - \alpha) \qquad (13)$$

$$\dot{\lambda} = \rho\lambda + z_G(w - s) \qquad (14)$$

$$\dot{G} = -(1 - \alpha)w + s + R \qquad (15)$$

From conditions (13) to (15), the behaviour of w along the optimal path can be derived:

$$\dot{w} = \frac{\rho\pi_w + z_G\left(\alpha s + R\right)}{\pi_{ww}} \qquad (16)$$

Considering cost function properties and concavity of $y(.)$, the $\dot{w} = 0$ locus has a positive slope:

$$w_G\big|_{\dot{w}=0} = -\frac{-\rho z_G + z_{GG}\left(\alpha s + R\right)}{\rho\pi_{ww}} > 0$$

Thus the steady state will be a saddle point. The $\dot{w} = 0$ locus may be convex or concave, depending on the signs of π_{www} and z_{GGG}, since:

$$w_{GG}\big|_{\dot{w}=0} = -\frac{\left[-\rho z_{GG} + z_{GGG}\left(\alpha s + R\right)\right]\rho\pi_{ww}}{\left(\rho\pi_{ww}\right)^2} + \frac{\rho\pi_{www}w_G\left[-\rho z_G + z_{GG}\left(\alpha s + R\right)\right]}{\left(\rho\pi_{ww}\right)^2}$$

The case of linear pumping costs and convex marginal benefits for water use ($\pi_{www} > 0$) provides an example of a convex $\dot{w} = 0$ locus. A phase diagram of the system might look like that of Figure 1. There is a stable arm that leads to the steady state equilibrium. For a given G_0, the chosen level of w_0 must be on that stable arm, so that the optimal path will converge to the steady state.

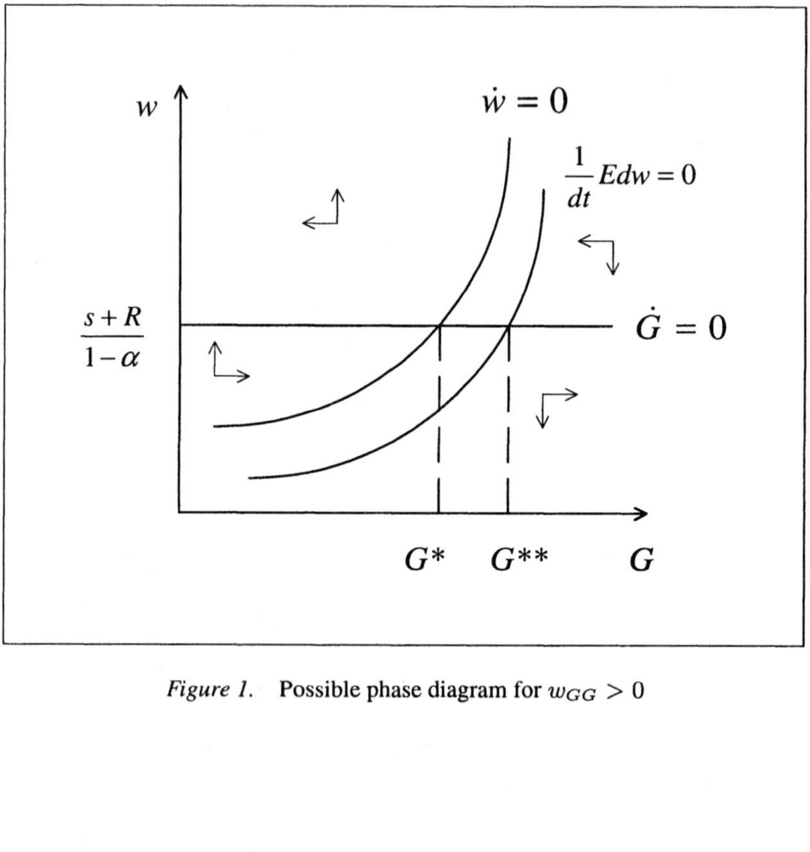

Figure 1. Possible phase diagram for $w_{GG} > 0$

3.2 Uncertainty in hydrological parameters

There are several ways in which uncertainty could affect the problem of groundwater extraction. One of them, as noted in section 2.1, is through surface water variability. When surface water and groundwater have different quality levels, these can also fluctuate, depending on weather conditions or imperfectly known pollution processes (some references to stochastic pollution processes can be found in Kampas and White, 2000; Shortle and Dunn, 1986; and Xepapadeas, 1992). Hence, none of the three hydrological parameters, s, C^s and C^g will generally be known with certainty, so that a stochastic setting in the decision problem may be more adequate.

It is assumed that the current realization of all parameters is known, although their future increments are stochastic, according to the following:

$$ds = \sigma_1 s d\omega_1 \tag{17}$$
$$dC^s = \sigma_2 C^s d\omega_2 \tag{18}$$
$$dC^g = \sigma_3 C^g d\omega_3 \tag{19}$$

where ω_1, ω_2, ω_3 are brownian motions with correlation coefficients given by ρ_{12}, ρ_{13} and ρ_{23}, respectively.[9] The specific type of stochastic behaviour chosen for the hydrological parameters (as geometric brownian motions without drift) implies that each of them is lognormally distributed, taking only positive values and with constant expected value (equal to its initial value). Similar assumptions are used in Fisher and Rubio (1997) for a hydrological parameter like s. If a drift component was relevant in shaping the behaviour of s, C^s or C^g, it would have to be incorporated in the above equations (see also Roseta-Palma, 2000).

The expected present value of total discounted profit is similar to that of problem (12). Thus the optimal value function will be:

$$V(G, s, C^s, C^g) =$$
$$\max_{\{w(t)\}} E_0 \int_0^\infty [y(w(t), C(t)) - z(G(t))(w(t) - s)]\, e^{-\rho t} dt \tag{20}$$

The associated Bellman equation is:

$$\rho V(.) = \max_{\{w\}} \; y(w, C) - z(G)(w - s) + \tfrac{1}{dt} EdV \tag{21}$$

Label as x the set of variables $x = \{G, s, C^s, C^g\}$, and define the Jacobian as V_x, the Hessian as V_{xx}, the transition vector as $T_x = (dG, ds, dC^s, dC^g)$, as well as $\sigma = (\sigma_1 s, \sigma_2 C^s, \sigma_3 C^g)$. Then, using Ito's Lemma:

$$dV = V_x T_x' + \frac{1}{2} T_x V_{xx} T_x' \tag{22}$$

Expanding terms and taking the expected value of dV as $dt \to 0$ yields:[10]

$$\begin{aligned}\frac{1}{dt} EdV = {} & V_G\left(-(1-\alpha)w + s + R\right) + \\ & \frac{1}{2}\left[\sigma_1^2 s^2 V_{ss} + \sigma_2^2 (C^s)^2 V_{C^s C^s} + \sigma_3^2 (C^g)^2 V_{C^g C^g}\right] + \\ & \sigma_1\sigma_2 s C^s V_{sC^s}\rho_{12} + \sigma_1\sigma_3 s C^g V_{sC^g}\rho_{13} + \sigma_2\sigma_3 C^s C^g V_{C^s C^g}\rho_{23}\end{aligned} \tag{23}$$

Or, in more compact notation, with $\widehat{V_{\chi\chi}} = [\rho_{ij} V_{ij}]$ for $i, j = s, C^s, C^g$ (ie. χ refers to elements of x except G):[11]

$$\frac{1}{dt} EdV = V_G\left(-(1-\alpha)w + s + R\right) + \frac{1}{2}\sigma \widehat{V_{\chi\chi}} \sigma' \tag{24}$$

Replacing expression (24) in equation (21) and undertaking the maximization yields:

$$\pi_w = (1-\alpha)V_G \tag{25}$$

Differentiating equation (21) with respect to G at the optimal value of w :

$$\begin{aligned}\rho V_G &= -z_G(w-s) + V_{GG}dG + \\ &\frac{1}{2}\left[\sigma_1^2 s^2 V_{ssG} + \sigma_2^2 (C^s)^2 V_{C^sC^sG} + \sigma_3^2 (C^g)^2 V_{C^gC^gG}\right] + \\ &\sigma_1\sigma_2 s C^s V_{sC^sG}\rho_{12} + \sigma_1\sigma_3 s C^g V_{sC^gG}\rho_{13} + \sigma_2\sigma_3 C^s C^g V_{C^sC^gG}\rho_{23}\end{aligned} \tag{26}$$

Now, considering that $V_G = V_G(G, s, C^s, C^g)$, and using Ito's lemma to obtain $\frac{1}{dt}EdV_G$, this expression reduces to:

$$\rho V_G = -z_G(w-s) + \frac{1}{dt}EdV_G \tag{27}$$

Equations (25) and (27) are the counterparts for the stochastic problem of equations (13) and (14), and they can be used to find the stochastic equivalent of the optimal path for water consumption, equation (16). Using similar notation as above, but defining $X = \{w, x\}$ (ie. w and all elements of x), and noting that $\pi_w = \pi_w(w, G, s, C^s, C^g)$:

$$d\pi_w = \pi_{wX}T_X' + \frac{1}{2}T_X\pi_{wXX}T_X' \tag{28}$$

Before expanding equation (28), the expressions for dw and $(dw)^2$ must be developed. Along the optimal path, $w = w(G, s, C^s, C^g)$, so that:

$$dw = w_x T_x' + \frac{1}{2}T_x w_{xx} T_x' \tag{29}$$

As for $(dw)^2$, it is greatly simplified by recalling that all terms of order higher than dt can be discarded, leaving:

$$(dw)^2 = T_x w_x' w_x T_x' \tag{30}$$

Equation (28) can now be rewritten, in expected value form (considering infinitesimal dt):[12]

$$\begin{aligned}\frac{1}{dt}Ed\pi_w &= \pi_{ww}\frac{1}{dt}Edw + \pi_{wG}\frac{dG}{dt} + \frac{1}{2}\left\{\pi_{www}\left[\widehat{\sigma w_X' w_X \sigma'}\right] + \widehat{\sigma \pi_{wXX}\sigma'}\right\} \\ &+ \pi_{wwX}\left[w_X\widehat{\sigma'\sigma}\right]\end{aligned} \tag{31}$$

where, as before, a matrix with a hat means that each of its elements appears multiplied by the appropriate correlation coefficient. Define:

$$A = \frac{1}{2}\left\{\pi_{www}\left[\widehat{\sigma w_{\chi}' w_{\chi}\sigma'}\right] + \widehat{\sigma\pi_{w\chi\chi}\sigma'}\right\} + \pi_{ww\chi}\left[w_{\chi}\widehat{\sigma'\sigma}\right] \tag{32}$$

Using conditions (25) and (27):

$$\rho\frac{\pi_w}{1-\alpha} = -z_G(w-s) + \frac{\frac{1}{dt}Ed\pi_w}{1-\alpha} \tag{33}$$

Replacing $\frac{1}{dt}Ed\pi_w$ with the expression obtained in equation (31), substituting $\pi_{wG}dG$ and reorganizing terms:

$$\frac{1}{dt}Edw = \frac{\rho\pi_w + z_G(\alpha s + R) - A}{\pi_{ww}} \tag{34}$$

This expression can be compared with its deterministic counterpart, equation (16). The stochastic steady state equilibrium, if it exists in the sense of convergence to a distribution for w and G, will satisfy $\frac{1}{dt}Edw = \frac{1}{dt}EdG = 0$.Thus the sign of A will determine whether expected steady state water stock is greater or smaller than in the deterministic case. If $A > 0$ then the $\frac{1}{dt}Edw = 0$ locus is below the $\dot{w} = 0$ locus and expected water stock is greater with uncertainty, as can be seen in the phase diagram of Figure 2. If $A < 0$ the opposite occurs. Unlike the case presented by Fisher and Rubio (1997), it is not sufficient to have convex marginal benefits to ensure a clear result, since A has a number of additional terms with generally unknown signs. Note that the term $\frac{1}{2}\left\{\pi_{www}\left[\widehat{\sigma w_{\chi}' w_{\chi}\sigma'}\right]\right\}$ is positive if $\pi_{www} > 0$ because all terms in $\widehat{\sigma w_{\chi}' w_{\chi}\sigma'}$ are positive.

3.2.1 Surface water variability

Since the derivation of equation (34) in the general uncertain case above is rather abstract, it might be useful to look at the case of only one uncertain variable so that the meaning of those extra terms in A is clarified. When surface water is variable (with increments described by equation (17) as before), the expanded version of equation (24) is simply:

$$\frac{1}{dt}EdV = V_G\left(-(1-\alpha)w + s + R\right) + \frac{1}{2}\sigma_1^2 s^2 V_{ss} \tag{35}$$

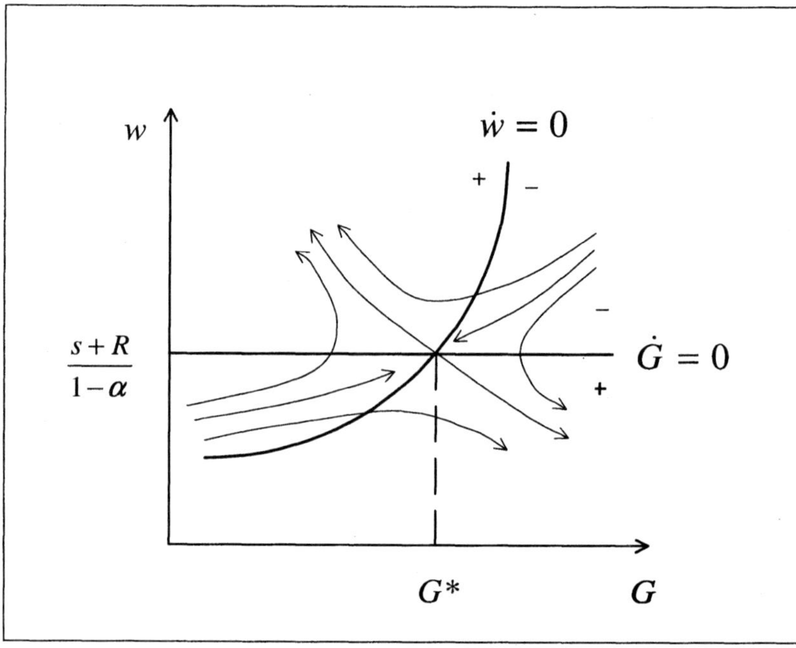

Figure 2. Phase diagram for $A > 0$

As for $d\pi_w$ and terms in dw, expressions (28), (29) and (30) reduce to:

$$\begin{aligned} d\pi_w &= \pi_{ww}dw + \pi_{wG}dG + \pi_{ws}ds \\ &\quad + \frac{1}{2}\left[\pi_{www}(dw)^2 + \pi_{wss}(ds)^2\right] + \pi_{wws}dwds \end{aligned} \tag{36}$$

$$dw = w_s ds + w_G dG + \frac{1}{2}w_{ss}\left(ds\right)^2 \tag{37}$$

$$(dw)^2 = \left(w_s\right)^2 \sigma_1^2 s^2 dt \tag{38}$$

So that the expression for the optimal expected motion of water consumption is:

$$\frac{1}{dt}Edw = \frac{\rho\pi_w + z_G(\alpha s + R) - \sigma_1^2 s^2 \left\{\frac{1}{2}\left[\pi_{www}(w_s)^2 + \pi_{wss}\right] + \pi_{wws}w_s\right\}}{\pi_{ww}} \tag{39}$$

This equation corresponds to equation (34) except shocks exist only in s. It is clear now that the additional terms in A result directly from the inclusion of surface water in the production function through weighted-average concentration, since instantaneous profit is no longer linear in s. Thus the cross-derivatives of π_w with respect to s do not disappear.

The effects of increasing surface water variability (ie. increasing σ_1) on stock size can also be derived analytically, for a given level of surface water. At the steady state, $\frac{1}{dt}Edw = \frac{1}{dt}EdG = 0$, so that $0 = Es + R - (1-\alpha)Ew$, which implies:

$$\overline{w} \equiv Ew = \frac{\overline{s} + R}{1 - \alpha} \tag{40}$$

Furthermore, evaluating all derivatives at $\overline{w}$ and $\overline{s}$:

$$\rho\pi_w + z_G(\alpha\overline{s} + R) - \sigma_1^2\overline{s}^2\left\{\frac{1}{2}\left[\pi_{www}(w_s)^2 + \pi_{wss}\right] + \pi_{wws}w_s\right\} = 0 \tag{41}$$

None of the terms in A depends on G, so the total differential of (41) is:

$$\left[-\rho z_G + z_{GG}(\alpha\overline{s} + R)\right] dG - \overline{s}^2\left\{\frac{1}{2}\left[\pi_{www}(w_s)^2 + \pi_{wss}\right] + \pi_{wws}w_s\right\} d\sigma_1^2 = 0 \tag{42}$$

From this expression it is clear that:

$$\frac{dG}{d\sigma_1^2} = \frac{\overline{s}^2\left\{\frac{1}{2}\left[\pi_{www}(w_s)^2 + \pi_{wss}\right] + \pi_{wws}w_s\right\}}{\left[-\rho z_G + z_{GG}(\alpha\overline{s} + R)\right]} \tag{43}$$

The denominator is positive under cost convexity, so that the sign of $\frac{dG}{d\sigma_1^2}$ will be the same as the sign of A. If $A > 0$, increasing the variance results in a higher groundwater stock, which is consistent with the result shown in Figure 2, since in that case variance is going from zero to a positive value.

4. CONCLUSION

A truly integrated approach to water management must embody not only quantity-quality interactions but also conjunctive use of surface water and groundwater. This paper is an attempt at analysing optimal choices when both aspects of water systems are considered, emphasizing the economic implications of conjunctive use when the quality of the water varies according to the source.

Water productivity depends on its quality. When different types of water are mixed, the relevant pollutant concentration is a weighted average of individual concentration levels. This simple fact alters a well established result in the conjunctive use literature, which was that for different levels of surface water endowments, groundwater would be pumped so as to keep a given optimal level of total water consumption, as that level was invariant with respect to surface water realizations. Now the optimal level of water consumption will no longer remain the same, and performing comparative statics shows that its reaction to parameter variations will always depend on the difference between surface water quality and groundwater quality.

Another aspect that has rightly received attention in the water management literature regards the effect of uncertainty on optimal choices. In this chapter uncertainty in hydrological parameters was modelled in a dynamic setting through their description as geometric brownian motions, and the impact of such uncertainty on the evolution of water consumption and on optimal steady state stock was described, although no general results can be obtained without specifying a production function.

There are two aspects that have been treated in the literature and were not incorporated in the present analysis. One deals with the choice of optimal storage capacity in the context of ground or surface water stocks (see Tsur, 1990; Fisher and Rubio, 1997). Another deals with models where surface water does not have to be entirely consumed, so that surface water used and groundwater pumped are actually two different, albeit related, choices. With uncertainty in quality parameters, each type of water has different risk and return. The conjunctive use problem could then be viewed as a choice of optimal portfolio mix. These are areas for future research.

Notes

1. The expected sign of the second derivative on concentration, y_{CC}, depends on whether additional pollution is more harmful for small values of concentration or for large ones. This is an empirical question. See Letey and Dinar (1986), which contains a number of estimated agricultural production functions when the quality problem is salinity.

2. If s was costly or $C^g - C^s < 0$, then the choice of s might become endogenous, although there would still be an exogenous maximum available amount of surface water (this makes sense for certain types of surface water, such as stream flows and lakes, and not for others, such as precipitation). The water management problem would become:

$$\begin{array}{ll} \max_{w,s} & y(w, C^s \frac{s}{w} + C^g \frac{w-s}{w}) - z(w-s) \\ s.t. & s \le s^{\max} \end{array}$$

3. The first order condition for the model without quality is the same as for the model with constant quality $C^g = C^s = C$, ie. $y_w = z$. This expression does not depend on s.

4. If only groundwater is used ($s = 0$) this result also holds, as expected.

5. Note that without quality considerations $y(\overline{s}, C^s) - E\{y(s, C^s)\} = y(\overline{s}, 0) - E\{y(s, 0)\}$; since only surface water is being used in either case, the $y(.)$ simply shifts down when $C^s > 0$.

6. As for $\Pi_{C^g} = y_C \frac{w-s}{w}$ and $\Pi_{C^s} = y_C \frac{s}{w}$, they are both negative, as expected.

7. The one stock/one flow model can also be used in the absence of groundwater, whenever there are alternative sources of surface water of which at least one is a stock.

8. t subscripts have been dropped for ease of exposition.

9. The increments of brownian motions have mean zero and variance dt (thus $E(d\omega_i) = 0, E(d\omega_i)^2 = dt$, $E(d\omega_i d\omega_j) = \rho_{ij} dt, i, j = 1, 2, 3$). For an introduction to stochastic processes, see Dixit and Pindyck (1994) Ch.3.

10. In going from (22) to (24), all terms in $d\omega_i$ disappear as their expected value is zero. Terms of order dt are kept, whereas terms in dt of any order higher than one go to zero.

11. Note that second derivatives with respect to G are absent, since the transition for stock does not have a variance term. Also note that $\rho_{ii} = 1$.

12. Note that π_{www} is a scalar, whereas π_{wwy} is a 3×1 vector and π_{wyy} is a 3×3 matrix.

References

Burt, O.R., 1967. 'Temporal allocation of groundwater', *Water Resources Research*, 3, 45–57.

Dinar, A., 1994. 'Impact of energy cost and water resource availability on agriculture and groundwater quality in california', *Resource and Energy Economics*, 16, 47–66.

Dinar, A. and A. Xepapadeas, 1998. 'Regulating water quantity and quality in irrigated agriculture', *Journal of Environmental Management*, 54, 273–289.

Dixit, A.K. and R.S. Pindyck, 1994. *Investment under Uncertainty*, Princeton University Press, Princeton, New Jersey.

Fisher, A. and S. Rubio, 1997. 'Adjusting to climate change: Implications of increased variability and asymmetric adjustment costs for investment in water reserves', *Journal of Environmental Economics and Management*, 34, 207–227.

INAG, 2001. Plano nacional da água, Technical report, Ministério do Ambiente e Ordenamento do Território.

Kampas, A and B. White, 2000. 'Instrument choice for controlling stochastic nitrate pollution', Conference of the EAERE, Crete.

Knapp, K. and L. Olson, 1995. 'The economics of conjunctive groundwater management with stochastic surface supplies', *Journal of Environmental Economics and Management*, 28, 340–356.

Letey, J. and A. Dinar, 1986. 'Simulated crop-water production functions for several crops when irrigated with saline waters', *Hilgardia*, 54, 1–33.

Provencher, B., 1995. 'Issues in the conjunctive use of surface water and groundwater' in *Handbook of Environmental Economics*, edited by D. Bromley, Blackwell.

Provencher, B. and O. Burt, 1994. 'A private property rights regime for the commons: The case for groundwater', *American Journal of Agricultural Economics*, 76, 875–888.

Roseta-Palma, C., 2000. *Essays in Natural Resource Economics*, PhD thesis, Faculdade de Economia da Universidade Nova de Lisboa.

Rubio, S., 1992. 'Adjustment costs, uncertainty, and the theory of investment: The case of non-renewable natural resources', *Journal of Environmental Economics and Management*, 23, 139–153.

Rubio, S. and J. Castro, 1996. 'Long-run groundwater reserves under uncertainty', *Investigaciones Economicas*, XX(1), 71–88.

Shortle, J.S. and J.W. Dunn, 1986. 'The relative efficiency of agricultural source water pollution control policies', *American Journal of Agricultural Economics*, 68, 668–677.

Tsur, Y., 1990. 'The stabilization role of groundwater when surface water supplies are uncertain: The implications for groundwater development', *Water Resources Research*, 26(5), 811–818.

Tsur, Y., 1991. 'Managing drainage problems in a conjunctive ground and surface water system' in *The Economics and Management of Water and Drainage in Agriculture*, edited by A. Dinar and D. Zilberman, pages 617–636, Kluwer Academic.

Tsur, Y., 1997. 'The economics of conjunctive ground and surface water irrigation systems : Basic principles and empirical evidence from southern California' in *Decentralization and Coordination of Water Resource Management*, edited by D. Parker and Y. Tsur, pages 339–361, Kluwer Academic Publishers.

Tsur, Y. and Graham-Tomasi, T., 1991. 'The buffer value of groundwater with stochastic surface water supplies', *Journal of Environmental Economics and Management*, 21, 201–224.

Tsur, Y. and Zemel, A., 1997. 'On event uncertainty and renewable resource management' in *Decentralization and Coordination of Water Resource Management*, edited by D. Parker and Y. Tsur, Kluwer Academic Publishers.

Xepapadeas, A., 1992. 'Environmental policy design and dynamic nonpoint-source pollution', *Journal of Environmental Economics and Management*, 23, 22–39.

Chapter 12

THE IMPACT OF RECOVERING IRRIGATION WATER LOSSES ON THE CHOICE OF IRRIGATION TECHNOLOGY WITH HETEROGENEOUS LAND QUALITY AND DIFFERENT CROPS*

Petra J.G.J. Hellegers
Agricultural Economics Research Institute (LEI), The Netherlands

1. INTRODUCTION

Rapid population growth often results in increased water scarcity and consequently in an interest in improving the productivity of irrigation water use, since irrigated agriculture is by far the largest consumer of water. Worldwide, about 72% of the water extracted is used for the production of food and fibres. Irrigated agriculture provides about 40% of the world's food supply, but occupies only 17% of the cultivated area (OECD, 2002). Concern about improving the irrigation effectiveness in this sector has therefore been widely reflected in the water economics literature (Caswell and Zilberman, 1985 & 1986; Shah et al., 1993; Shah and Zilberman, 1991).

An increase in the irrigation effectiveness is often put forward as 'the solution' to the problem of reducing irrigation water use and losses. The adoption of modern irrigation technologies is therefore often encouraged. Whether such adoption is desirable in all cases is examined in this chapter. It is often argued that adoption is not interesting when losses fulfil leaching requirements or when losses are recovered through surface run-off returning to the hydrological cycle, through percolation to groundwater or indirectly through capillary rise. In such cases it seems interesting to reduce losses only to save pumping costs or to avoid environmental quality degradation or

* The author would like to thank Renan Goetz and Dolors Berga for their helpful comments.

water-logging. Applied irrigation water that is not utilised by the crop is, for instance, lost in saline groundwater areas. It therefore seems desirable to increase the irrigation effectiveness in saline areas, but not in fresh groundwater areas where losses can be recovered. Finally, it is not clear whether an increase in the irrigation effectiveness will really reduce water use, since effective water use becomes cheaper.

The aim is therefore to study in which cases an increase in the irrigation effectiveness by adopting modern irrigation technologies is desirable. To study the impact of land quality on the choice of irrigation technology, the technology choice model of Caswell and Zilberman (1986) presented in Section 2 is employed. This conceptual model will be used to clarify whether the modern irrigation technology saves water or increases water demand.

As an extension of existing work, in Section 3 the recovery of irrigation water losses is introduced. The extended model will be used to study the impact of the recovery of irrigation water losses on the choice of irrigation technology with heterogeneous land quality. Moreover, the possibility to grow different crops on land with different quality is introduced. This extension can be considered a contribution to the literature, since crop choice has yet not been modelled as an endogenous variable in the technology choice model of Caswell and Zilberman (1986). Another contribution to the literature is the introduction of decreasing costs of investment in new technology, whereas investment costs in existing models do not vary with the size of the activity. Section 4 provides an insight into private versus social incentives to adopt modern irrigation technologies, if the costs of externalities due to irrigation are not internalised. Section 5 presents the summary and conclusions.

2. THEORETICAL MODEL

The technology choice model of Caswell and Zilberman (1986) illustrates the interaction between economic and biophysical variables in determining the rate of water use and the choice of irrigation technology. Farmers may employ various irrigation technologies that determine which fraction of applied water (a) is actually utilised by the crop; this is often referred to as the irrigation effectiveness (h) of the technology. The amount of applied water that is actually used by the crop is referred to as effective water (e)[1]. The irrigation effectiveness h_i of technology i is the ratio between the effective and the applied water:

$$h_i(\alpha)=e_i(\alpha)/a_i(\alpha) \quad (1)$$

where α denotes a land quality index. It is defined in terms of the ability of the land to store water (which can be utilised by the crop over time), which depends on soil permeability, water-holding capacity and slope of the land. The land quality index α ranges from 0 to 1 to indicate poor to good land quality. Steep land and sandy soils correspond to a low value of α, while flat and heavier land correspond to a high value of α. Irrigation effectiveness is higher on heavier clay soils (high-quality land) than on sandy soils (low-quality land), where applied water is very poorly retained. The irrigation effectiveness increases at a decreasing rate with land quality, namely $h'_i(\alpha) > 0$ and $h''_i(\alpha) < 0$, implying that the gain in irrigation effectiveness from a technology switch declines as land quality improves (Shah et al., 1995).

To illustrate how land quality influences farmers' choice of irrigation technology and water use, the following model is used. Output per hectare *y* is, ceteris paribus, given by $y = f(e)$, with $f'(e) > 0$ and $f''(e) < 0$, that is, $f(e)$ is an increasing and concave neo-classical production function. Two technologies are considered: a traditional one $i = 1$ and modern one $i = 2$. It is assumed that the modern technology has a higher irrigation effectiveness than the traditional technology: $1 > h_2 > h_1 > 0$. The modern technology produces the same maximum output per hectare as the traditional technology $f(h_1a_1) = f(h_2a_2)$ *but requires less water* ($a_2 < a_1$). However, it requires higher investments per hectare $k_2 > k_1$.

Quasi-rent Π_i per hectare is equal to agricultural output price *P* times output per hectare, minus the price of applied water *w* times the quantity of water applied a_i and the cost of technology k_i per hectare for each technology *i*. The profit-maximising choice of water application and irrigation technology is solved via a two-stage procedure. First, the optimal amount of water for each technology is chosen, and then the most profitable irrigation technology is selected. In other words, the maximal competitive quasi-rent Π^* is obtained by determining initially the optimal level of applied water a_i^* for each technology *i*, and thereafter the technology by evaluating

$$\Pi^* = \max_{i=1,2}\left\{\max_{a_i}\{Pf(h_i(\alpha)a_i) - wa_i - k_i\}, \text{for i=1,2}\right\} \tag{2}$$

The quasi-rent difference between the two technologies can be written as:

$$\Delta\Pi = P\Delta y - w\Delta a - \Delta k \tag{3}$$

The modern technology will be selected in the case where the increase in quasi-rents from higher yields or lower water cost offsets the higher investment cost. These results indicate that the adoption of modern irrigation technology will increase with increasing output or increasing water prices (Zilberman and Lipper, 1999). In other words, a higher price of water provides incentives to adopt the modern technology. The quasi-rent technology i will be maximal when the value of the marginal product of effective water is equal to the price of effective water, that is

$$Pf'(h_i a_i) = \frac{w}{h_i}, \text{ for } i = 1, 2 \tag{4}$$

As the price of effective water use w/h_i is lower under the modern than under the traditional technology (since $h_2 > h_1$), it is likely that *effective water* use will increase due to the adoption of modern technology. Whether *applied water* use will increase due to a switch in irrigation technology depends on additional properties of the production function. The elasticity of marginal productivity of effective water (EMP) measures how responsive the crop is to further irrigation. It is defined as:

$$\varepsilon_i(e) = -f''(e_i)e_i / f'(e_i) \tag{5}$$

According to Caswell and Zilberman (1986), optimal applied water use is lower with the modern technology if EMP is greater than 1, which is consistent with empirical evidence concerning this relationship for irrigation water. Caswell et al. (1990) show that adoption is always output-increasing but that it is input-saving only if EMP is greater than 1. Under most conditions, adoption results in both a decrease in applied water use and an increase in yields.

3. RECOVERY OF IRRIGATION WATER LOSSES

An increase in the irrigation effectiveness by adopting modern irrigation technology is often put forward as 'a solution' to the problem of reducing irrigation water use and irrigation water losses. In this section, it is analysed whether this result still holds when water losses are recovered. The model presented above is used to study the impact of recovery of irrigation water losses on the choice of irrigation technology with heterogeneous land quality and different crops.

In some cases the residual quantity of applied water (i.e. the water not utilised by the crop) is recovered through surface run-off returning to the hydrological cycle, through percolation to groundwater or indirectly through capillary rise. If this is the case, it may not be justified to charge resource costs for recovered losses. The price of applied water w (€/m^3) is therefore split into two parameters that indicate pumping costs u (€/m^3) and resource costs v (€/m^3), respectively.

$$wa_i = ua_i + vh_i(\alpha)a_i + v(1 - h_i(\alpha))a_i \quad (6)$$

If all irrigation water losses are recovered, only the pumping costs of applied water ua_i and the resource costs of effective water use $vh_i(\alpha)a_i$, should be paid by the farmer. Hence the maximal quasi-rent for technology i is determined by:

$$\Pi_i^* = \max_{a_i}\{Pf(h_i(\alpha)a_i) - ua_i - vh_i(\alpha)a_i - k_i\}, \text{ for } i = 1,2 \quad (7)$$

Comparing the quasi-rents without recovery (inner part of equation 2) and the quasi-rent in the case of recovery (equation 7), demonstrates that recovery increases the quasi-rent by $v(1\text{-}h_i(\alpha))a_i$*. This reduction in resource costs increases the profitability of the traditional technology relative to the modern technology, making the adoption of modern technology less likely. In other words, it is less interesting to reduce water losses when they are recovered.

In some cases, however, the share of applied water which is not utilised by the crop may be a source of environmental pollution[2]. As an example, the nitrate concentration in recharge and irrigation flows are considered. When recharge flows are of a lower quality than irrigation flows, the costs of such polluting losses should be taken into account. The difference between the nitrate concentration in recharge flows[3], N^{R} (measured in g/m^3) and the nitrate concentration in irrigation water, N^{I} (also measured in g/m^3) is used here as an indicator for a change in water quality. Thus, the maximal quasi-rent for technology i is given by

$$\Pi_i^* = \max_{a_i}\{Pf(h_i(\alpha)a_i) - ua_i - vh_i(\alpha)a_i - k_i - \tau(1 - h_i(\alpha))a_i(N^{\mathrm{R}} - N^{\mathrm{I}})\} \quad (8)$$

If $N^{\mathrm{R}} > N^{\mathrm{I}}$, the imposition of a pollution tax τ (€/g) on an increase in nitrate concentration will augment the profitability of the modern technology relative to the traditional technology. In other words, modern technology adoption is more likely when the costs of environmental pollution are taken into account. If $N^{\mathrm{R}} < N^{\mathrm{I}}$, a subsidy will be granted depending on the decrease in the nitrate concentration and on the amount of lost water. In this case it will not be interesting to reduce water losses. The impact of return flows on the quality of the stock depends on the dilution effect (Hellegers et al., 2001). Larger stocks tend to slow down changes in quality. However, the size of the stock is not considered in this chapter.

Thus, recovery of water losses reduces resource costs by $v(1-h_i(\alpha))a_i$, but imposes additional costs of $\tau(1-h_i(\alpha))a_i(N^R - N^I)$, if $N^R > N^I$. Consequently, recovery increases the profitability by $(v - \tau(N^R - N^I))(1-h_i(\alpha))a_i$. In this case, the profitability of modern technology will increase relative to the profitability of traditional technology if $v < \tau(N^R - N^I)$, and thus modern technology adoption is more likely. It is less likely if $v > \tau(N^R - N^I)$. In other words, technology adoption depends on the size of the tax τ relative to the resource costs v.

It is important to note that farmers often only cover part of the resource and pollution costs, due to the public nature of these costs. They frequently pay only for the pumping costs, which leads to the following expression for the determination of the maximal quasi-rent for technology i:

$$\Pi_i^* = \max_{a_i}\{Pf(h_i(\alpha)a_i) - ua_i - k_i\} \text{ for } i = 1,2 \tag{9}$$

Hence, farmers may not benefit from a reduction in resource and pollution costs as a result of modern technology adoption. Farmers benefit in that case only from a saving on pumping costs under modern technologies, since adoption results under most conditions in a decrease in applied water use (Section 2). Private incentives for adoption may therefore be insufficient to achieve socially desired levels of adoption, which necessitates government intervention. The impact of cost recovery on technology choice is discussed in more detail in Section 4.

3.1 Technology adoption decisions with heterogeneous land quality

The previous analysis showed that the impact of recovery of water losses on the quasi-rent depends on the irrigation effectiveness, which increases at a decreasing rate with land quality. As land quality improves, the output-increasing and the input-saving effects of modern technology decrease, such that the difference in effectiveness between the modern and the traditional technology continuously decreases. Given the higher fixed cost of the modern technology and smaller gains from adoption as land quality improves, modern technology will not be adopted on high-quality land (Shah et al., 1995).

Figure 1 shows the quasi-rent in form of solid lines as a function of land quality. On land where quality is below α_i^m (marginal land quality for each technology, at which quasi-rents are zero) agriculture is not profitable and land is left fallow. On land where quality is above α^s (switching land quality), the traditional technology is preferred. On land where quality is moderate (between $\alpha^m{}_2$ and α^s), the modern technology is preferred. The modern technology increases the quasi-rent on land between $\alpha^m{}_1$ and α^s, and makes land between α_2^m and α_1^m profitable, which implies that the modern technology augments land quality. By definition $\Pi_1(\alpha^s) = \Pi_2(\alpha^s)$, $\Pi_2(\alpha_2^m) = 0$ and $\Pi_1(\alpha_1^m) = 0$. Higher water prices increase α_i^m. Higher output prices decrease α_i^m, and increase α^s, such that they widen the range of land qualities using a modern technology.

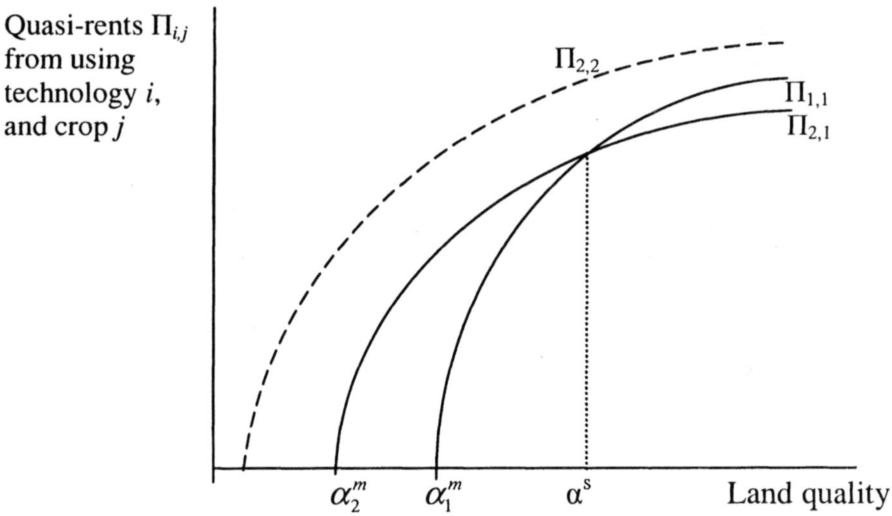

Figure 1. *Quasi-rents as a function of land quality for a traditional i=1 and modern technology i=2, and for two different crops j=1,2 under the modern technology (dashed line)*

Thus, the pattern of adoption of the modern technology on land of varying quality depends on how the relative monetary gain from adoption changes as the land quality varies. Therefore, the pattern of adoption in a particular region depends on the composition of the land quality in that region. Khanna et al. (2002) introduce a continuous land distribution function that represents the density of hectares of land with quality α within the region. If the share of land with good quality is relatively high, the adoption rate will be low.

3.2 Technology adoption decisions with different crops

So far the conceptual model allows farmers only to select the technology and the quantity of applied irrigation water for a given crop. In reality, however, farmers may choose simultaneously between both crops and technology. Dinar and Zilberman (1991) compare the profits of cotton and tomatoes to analyse crop choices under various output market prices.

One aspect, which has not been dealt with in the literature, is the fact that different land qualities allow the growing of different crops. As the modern irrigation technology augments land quality, adoption can partly eliminate differences in land quality. Thus, a switch in irrigation technology may induce a switch in crop choice or cropping intensity, as land quality improves. If such a switch in the cultivated crop or cropping intensity occurs, the quasi-rent function will increase even more. Hence, the quasi-rent $\Pi_{i,j}$ depends on the type of irrigation technology i and on cultivated crop j.

The impact of a change in the cultivated crop on the quasi-rent may be stronger than the effect of an increase in land quality. In other words, switching from crop $j = 1$ to $j = 2$ might change the quasi-rent of the modern technology such that it is always superior to the traditional technology; see the dashed line in Figure 1. It shows that the quasi-rent of crop $j = 2$ may exceed the quasi-rent of crop $j = 1$ even on high-quality land. For $j = 2$ there does not exist a switching land quality where $\Pi_{1,2}(\alpha^s) = \Pi_{2,2}(\alpha^s)$, since $\Pi_{2,2} > \Pi_{1,2}$ for all α.

The profit-maximising choice of applied water, technologies and crops is solved via a two-stage procedure. First, the optimal amount of water for each technology/crop combination is chosen, and then the most profitable technology/crop combination is selected. The maximum competitive quasi-rent $\Pi^*_{i,j}$ is obtained by solving for the following mathematical problem.

$$\Pi^*_{ij} = \max_{ij}\left\{\max_{a_{ij}}\left\{P_j f(h_{ij}(\alpha)a_{ij}) - wa_{ij} - k_i\right\}\right\}, \text{ for } i = 1,2 \text{ and for } j=1,2 \quad (10)$$

Whether a switch in cultivated crops will take place or not depends on, among others factors, the output market prices P_j and the current agricultural-environmental policies in effect. If crop choices change on a larger scale, output prices may change as well. Hence, a future extension of our work may want to consider endogenous rather than exogenous prices.

Decreasing costs of investment in new technologies

Adoption shows individual behaviour as a response to a new innovation. It can be depicted by a discrete choice variable, which shows whether or not the new technology is employed. Alternatively, adoption can be presented by a continuous variable which indicates to what extent the new technology is utilised, for instance measured by the number of hectares where the new technology is adopted. Figure 1 does not show to what extent the technology is adopted, whereas the size of the activity might be determining for the investment costs k_i per hectare of technology i. Costs of sprinkler equipment per hectare may, for instance, decrease with the number of hectares, $m_i = \sum_{j=1}^{j=2} m_{i,j}$, utilised, where $m_{i,j}$ reflects the number of hectares of each technology/crop combination.

If this is the case, the costs per hectare of each technology depend on the number of hectares irrigated with that technology. The resulting investment cost function for technology i, $k_i(m_i)$ $i = 1,2$, is assumed to be decreasing and convex, that is, $k'_i(m_i) < 0$ and $k''_i(m_i) < 0$. This implies that costs decrease at a declining rate with the number of hectares irrigated. If only one hectare is irrigated with the new technology, investment costs per hectare will be high and the quasi-rent will consequently be low. This implies that the number of hectares irrigated with the new technology has to exceed a certain threshold level, for instance $m_i > x$, before the new technology/crop combination will become the most profitable choice.

Total quasi-rent $\prod$ of the farm is equal to the sum of quasi-rent per hectare $\Pi_{i,j}$ times the number of hectares of that particular technology/crop combination: $\prod = \sum_{i=1}^{i=2} \sum_{j=1}^{j=2} \Pi_{i,j} m_{i.j}$

The maximal total quasi-rent $\prod^*$ of the farm can be obtained by determining simultaneously the optimal amount of water for each technology/crop combination and the optimal number of hectares on which to utilise that particular technology/crop combination:

$$\prod^*(a_{ij}, m_{i,j}) = \max_{a_{ij} m_{ij}} \sum_{i=1}^{i=2} \sum_{j=1}^{j=2} (p_j f(h_{ij}(\alpha) a_{ij}) - w a_{ij} - k_i(\sum_{j=1}^{j=2} m_{ij})) m_{ij} \quad (11)$$

The optimal number of hectares on which to utilise a particular technology/crop combination m_{ij} depends in turn on the land quality distribution within the region, that is, the share of land with a particular quality α in the entire region (Khanna et al., 2002).

4. PRIVATE VERSUS SOCIAL INCENTIVES FOR ADOPTION DUE TO COST RECOVERY

This section provides an insight into private versus social incentives to adopt modern irrigation technologies, if the costs of externalities due to irrigation are not internalised. Recovering irrigation water losses reduces resource costs, but may incur additional pollution costs. It is interesting to reduce both cost components especially on low-quality land. Due to the public nature of pollution costs, social gains from reducing these costs often exceed private gains.

Whether adoption is desirable and likely from a private point of view depends on who has to bear the costs of such externalities. When only a part of the costs is recovered in the price of water, farmers probably do not benefit from reducing pollution costs. Social gains from a switch in technology often exceed private gains. The farmer maximises the private quasi-rent (Equation 9) instead of the social quasi-rent (Equation 8), which gives fewer incentives for adoption in the presence of externalities. This stresses the need to internalise these costs in the price of water.

Transaction costs of charging for pollution costs may, however, be high. For instance, the monitoring of water quality is very costly. The transaction costs may even exceed the gains from modern technology adoption. It can therefore be more attractive to encourage adoption by means of a subsidy than by means of full cost recovery. Thus, whether technology adoption should be encouraged by means of a subsidy or by means of full cost recovery depends on the size of the transaction costs of charging for externalities.

While there are some private incentives to adopt modern irrigation technologies, these incentives may be insufficient to induce socially desired levels of adoption. If it is socially desirable to increase the irrigation effectiveness in order to reduce water use or water losses, the government may encourage technology adoption. The government can promote diffusion by engaging in extension and education activities, by subsidising all or part of the fixed costs of modern technologies or by regulating adoption (by setting a timetable for diffusion).

Technology diffusion shows aggregate behaviour, that is, the gradual process in which technology spreads through the economy. The technology diffusion curve shows the number of adopters as a function of time (Sunding and Zilberman, 2001). It tends to be S-shaped and different phases can be distinguished. There is an initial phase of slow adoption (introduction). Then there is drastic adoption (take-off). In the final phase, adoption is again slow (saturation). If the government promotes a new technology or the full costs are recovered, the technology will be adopted more readily, and the diffusion curve will shift to the left.

5. SUMMARY AND CONCLUSIONS

Regarding the impact of recovering irrigation water losses and cost recovery on the choice of irrigation technology with heterogeneous land quality for different crops, this chapter shows that understanding the heterogeneity of environments in which modern irrigation technologies may be utilised is essential for a complete analyses of adoption processes. Moreover, the chapter provides an insight into the desirability of modern technology adoption from a private and from a social perspective.

Often it is argued that it is not necessary to increase the irrigation effectiveness when losses are recovered or when losses fulfil leaching requirements, since these losses are not real losses. However, this chapter shows that it can be desirable to reduce losses – and even recovered losses – when recovered water is of poor quality.

Another benefit from a reduction in irrigation water losses is the saving in pumping costs, since adoption results under most conditions in a decrease in applied water use. Effective water use is likely to be higher after the switch in irrigation technology, due to the lower price of effective water under the modern technology. A reduction in applied water and an increase in effective water imply a reduction in irrigation water losses.

Given that the difference in irrigation effectiveness gains between the modern and the traditional technology is small on high-quality land, and given the higher fixed cost of installing modern technology, it is likely that modern technology will not be adopted on high-quality land. Thus, modern irrigation technology adoption is more desirable on low-quality land if losses are not recovered and if recovered water losses are of poor quality.

The analyses also shows that a switch in irrigation technology may induce a switch in crop choice, as the modern technology augments land quality. Finally, due to the public nature of resource and pollution costs, farmers often do not internalise all costs. As they do not benefit from reducing these costs, private incentives to adopt modern technologies may be insufficient to induce socially desired levels of adoption This may explain why governments sometimes subsidise modern technology adoption. Another option is to internalise all costs in the price of water, which provides more incentives to adopt the new technology.

For future research it might be worthwhile to consider crop choice as an endogenous variable and to include transactions cost and size-dependent investment cost to determine the exact effect of these additional factors in a numerical analyses.

Notes

1. For the sake of simplicity, effective water use is not a function of water quality in the analysis. Where effective water use is a function of water quality, low-quality water reduces effective water use.

2. Non-point source pollution cannot be observed directly and is stochastic due to random variation in environmental conditions. To concentrate on the economic analysis of technology adoption, the focus is on the deterministic portion of this pollution, that is, the expected level of pollution that is influenced by applied water use and technology choice (Khanna et al., 2002).

3. For simplicity, we assume that N^R does not depend on the size of recharge flows or land quality.

References

Caswell, M. and D. Zilberman, 1985. 'The choices of irrigation technologies in California', *American Journal of Agricultural Economics*, 67(2), 224-234.

Caswell, M. and D. Zilberman, 1986. 'The effects of well depth and land quality on the choice of irrigation technology', *American Journal of Agricultural Economics*, 68(4), 798-811.

Caswell, M., E. Lichtenberg and D. Zilberman, 1990. 'The effect of pricing policies on water conservation and drainage', *American Journal of Agricultural Economics*, 72(4), 883-890.

Dinar, A. and D. Zilberman, 1991. 'Effects of input quality and environmental conditions on selection of irrigation technologies' in *The economics and management of water and drainage in agriculture*, edited by A. Dinar and D. Zilberman, Kluwer Academic Publishers, Boston, pp. 229-249.

Hellegers, P., D. Zilberman and E. van Ierland, 2001. 'Dynamics of agricultural groundwater extraction', *Ecological Economics*, 37(2), 303-311.

Khanna, M., M. Isik and D. Zilberman, 2002. 'Cost-effectiveness of alternative green payment policies for conservation technology adoption with heterogeneous land quality', *Agricultural Economics*, 27(2), 157-174.

OECD, 2002. *Transition to full-cost pricing of irrigation water for agriculture in OECD countries.* OECD, Paris.

Shah, F. and D. Zilberman, 1991. 'Government policies to improve inter-temporal allocation of water in regions with drainage problems' in *The economics and management of water and drainage in agriculture*, edited by A. Dinar and D. Zilberman, Kluwer Academic Publishers, Boston, pp. 637-660.

Shah, F., D. Zilberman and U. Chakravorty, 1993. 'Water rights doctrines and technology adoption' in *The economics of rural organisation: Theory, Practice and Policy*, edited by K. Hoff. A. Braverman and Stiglitz, New York: Oxford University Press, pp. 479-499.

Shah, F., D. Zilberman and U. Chakravorty , 1995. 'Technology adoption in the presence of an exhaustible resource: the case of groundwater extraction'. *American Journal of Agricultural Economics*, 77, 291-299.

Sunding, D. and D. Zilberman, 2001. 'The agricultural innovation process: research and technology adoption in a changing agricultural sector' in *Handbook of agricultural and resource economics*, edited by B. Gardner and G. Rausser, Amsterdam: North Holland.

Zilberman, D. and L. Lipper, 1999. 'The economics of water use' in *Handbook of environmental and resource economics*, edited by J.C.J.M. van den Bergh, Edward Elgar, Cheltenham.

Chapter 13

PRECISION FARMING IN COTTON

David Zilberman
University of California, Berkeley, U.S.A.

Daniel Cohen-Vogel
Florida Legislature, OPPAGA, U.S.A.

Jeanne Reeves
Cotton Incorporated, U.S.A.

1. INTRODUCTION

Precision farming is a new category of agricultural technologies that adjusts application levels of agricultural inputs to accommodate variations within fields and also to climatic and other variations within seasons. The introduction of precision technologies in agriculture has been motivated by the high degree of variability of agro-ecological conditions within fields (National Research Council, 1997). Within fields, differences in slope, soil, plant size, etc., may warrant varying the application of water and other input over space. Farmers in a labor-intensive farming system adjust their input use to accommodate variation. Increases in mechanization of farming led to increased use of uniform application levels within fields with minimal adjustment to variability. The modern precision technologies rely on new developments in remote sensing, telecommunication, and computation technologies as well as impending agronomical knowledge. It requires extra investment in equipment and monitoring efforts and, in return, aims to increase yield, reduce environmental side effects of agricultural production, and reduce input use (NRC, 1997).

Some features of precision technologies have already been adopted in agriculture. Some of the benefits of drip irrigation, various sensors, and automated and computerized irrigation systems is the gain from added precision as they allow to adjust irrigation in response to changes in soil situations and climatic conditions. As information and communication technologies evolve, the potential for improved performance and the costs of advanced precision systems decline. Nanotechnology and ground-penetrating radar groundwave techniques (Grote et al., 2003) provide new opportunities for information gathering that will enhance precision in irrigation. In this paper we first analyze the main features of precision farming technologies and present some of the lessons from past adoption processes that are likely to apply to modern precision irrigation. We then introduce a modeling framework to assess the gains from adoption of these technologies and the conditions under which the adoption of the technology makes economics sense. This framework is general, and we will introduce a simulation based on California cotton information to assess the order of magnitude of the potential gains from adoption of precision irrigation technologies.

2. THE MAIN FEATURES AND CHALLENGES OF PRECISION FARMING TECHNOLOGIES

Most of the literature on farm management treats fields as homogeneous micro-production units, thus prescribing the uniform application rates at the field level. Of course, variability within a field has not been totally ignored. Operators of combines and equipment such as cultivators were challenged to adjust their equipment to changes in space. In general, however, when it comes to applications of inputs such as pesticides, seeds, or fertilizers, little attention has been given to the variations within fields. Yet, there is evidence that there are high levels of soil quality variations and other types of heterogeneity within fields. For example, yield in some areas within fields are documented to be up to three times higher than in other locations. The reasons for these differences can be permanent, as is the case with variations in soil conditions, e.g., differences in soil depths, chemical composition, etc. Thus, differences that are more transitory include, for example, variations in pest levels or pest infestation. Different types of spatial variability call for different treatments.

Labor-intensive production systems have been very effective in treatment of spatial variability. Workers weed out pests whenever they occur, and planting and fertilizer applications may be manually adjusted to variability in conditions. High labor intensity occurs in systems where the cost of labor is relatively low compared to the cost of the final product, e.g., in strawberry production in the United States or production of fruits and vegetables and other agricultural commodities in poor developing countries. However, machinery has been introduced to the vast majority of modern crop systems in developed countries, and labor use has become uneconomical. Thus, the challenge is to develop automated systems that take into account spatial variability.

Randomness of climatic conditions such as temperature, rainfall, frost, etc., is the major cause for yield variabilities. Agricultural production systems are designed to adjust to random weather events. Farmers protect against crop loss by modifying irrigation in response to heat waves, etc. Adjustment strategies that adapt to random weather conditions need much improvement. These strategies are geared more towards responding to drastic weather events than to slight variations. Notably, traditional irrigation schedules in many cases are designed far in advanced, and often both timing and application do not change much unless drastic events occur.

New precision technologies aim to use equipment and knowledge to develop mechanisms to respond to both spatial and climatic variability. With the labor-intensive agricultural production systems, workers are able to observe differences in conditions and react accordingly. Precision farming systems are challenged to replace the eyes, the brains, and the hands of these workers.

Conceptually, precision farming systems must perform three major functions. First, they need to have the monitoring capacity to observe and assess the phenomenon of interest. For example, systems that aim to control weeds have to be able to monitor their growth in the field. Then they need the decision-making capacity to determine the appropriate response for each situation. Third, they need to be able to perform what is needed at the right place at the right time. These three functions are not necessarily conducted by the same piece of equipment.

Actually, the beauty of precision farming is that it may be used for different types of farming machinery and equipment to perform different tasks. For example, in some cases remote sensing will provide updated maps of field conditions and, if the farmer detects a high level of infestation at a certain part of the field, then the equipment may be transported to tackle this problem. In other situations, yield monitors and field testing will generate a geographic information system (GIS) of soil productivity, and computer programs will determine what the appropriate input mix is for each location based on this map. Then specialized equipment will apply varying rates of chemicals to adjust to the conditions of different locations.

The design of precision farming systems is quite challenging because it requires interdisciplinary knowledge and takes advantage of findings in various fields. The degree of accuracy and quality of performance of different precision farming systems may be quite different. For example, the capacity to monitor has been significantly enhanced with remote sensing, GIS, and the geographic positioning systems (GPS). Nanotechnologies will further expand the capacity to monitor environmental conditions. It may also provide new means to modify input application levels. Although the accuracy of varying application equipment may be improving, the ability to compute optimal treatment to accommodate variations in circumstances is still limited. We do not have enough data and biological knowledge to quantitatively obtain good estimates of basic functions, for example, the relationship between applications of inputs in various environmental conditions and the resulting output. But one of the most fascinating features of precision farming systems is the ability to constantly learn and improve the accuracy of these estimates and, thus, the performance of the systems.

3. PRECISION FARMING AND IRRIGATION

Precision farming methods can be applied to enhance the productivity of various agricultural inputs. They are introduced to adjust the application of fertilizers and seeds in various soil conditions. In some cases precision farming can be used to apply more than one seed variety on a field to account for variability. They can be used to detect pest problems and attack them wherever they occur. However, the applications of precision farming technologies and their benefits are significantly different in irrigated versus rain-fed agricultural systems.

The introduction of irrigation affects precision farming in several ways. First, an irrigation system has an added applied input, water, and its application over space and time can be better controlled by precision farming. Moreover, water can be the vehicle through which other inputs can be applied. Technologies and systems to improve irrigation and water applications have the basic philosophy of precision farming, even though this term has not been used historically to describe these systems. The improvements in profitability and performance, resulting from the introduction of new irrigation management methods and technologies in California cotton, provide indications of the potential and problems of introducing precision farming in other areas and aspects of production besides irrigation.

The traditional irrigation systems in California relied on application technologies such as furrow, border, or flood irrigation. With these systems, timing and level of applications were often predetermined at the start of the season. This has been especially the case when farmers received their water from water districts. Due to the infrastructures of the districts, which mainly include fixed canals, managers needed to know who were growing what and where. Managers then design a predetermined water diversion schedule to meet farmers' needs given the constraints of the system. These plans are quite complex and allow a minimal level of flexibility.

One major feature of traditional irrigation technologies was their relatively poor irrigation efficiency. Irrigation efficiency is a fraction of applied water that is actually used by the crop. In locations with relatively level fields and heavy soils, the irrigation efficiency of traditional methods may be quite high, up to 80% to 90%. However, in areas where the soil is sandy or the hills are steep, irrigation efficiency may be .5 or less, especially if the canals are very long. On average, irrigation efficiency in California of traditional technologies in Central Valley is considered to be 0.6.

Studies (Gardner, 1995; Zilberman et al., 1994; Feinerman et al., 1983) have shown that adoption of traditional irrigation technologies made economic sense in locations where water was relatively abundant and when the basic principles of water allocation were dictated by the prior appropriation system. Under this system, users established water rights based on seniority of time and, if water rights were not utilized, they were lost. Under such conditions, growers do not have the incentive to consider systems that conserve water.

A key feature of traditional surface water systems was that the operation of the canal and individual farm irrigation systems was relatively low in cost. Water is cheap and the canal system is linear and simple and does not allow much flexibility and changes. The canals themselves were not always enforced and, at times, had leakage problems. The leakage was not viewed negatively, since they generated groundwater reservoirs that served several purposes. They provided reservoirs in situations of drought (Burt, 1964). Furthermore, some of the groundwater was used by other users, for example, municipalities.

Expansion of irrigated acreage without parallel expansions in water availability, combined with occasional periods of drought, have provided an impetus to develop and adopt modern irrigation technologies. Furthermore, a large number of farmers pump groundwater for irrigation, and they have more flexibility since they control the time of pumping. They also pay for the pumping cost, which may be substantial for deep wells. Thus, they have better incentives to adopt modern technologies. Caswell and Zilberman (1985) found that the main variables that prevent adoption of irrigation technologies, such as sprinkler and drip irrigation, are reliance on surface water or shallow water levels when groundwater was used.

While traditional irrigation technologies are gravitational (gravity is the main force that distributes the water), with modern irrigation technologies, there is extra investment in equipment for water applications. Modern irrigation technologies include various types of sprinklers, drip, low energy precision application (LEPA), and low-pressure, center-pivot irrigation. The ability to increase precision with modern irrigation technologies varies substantially by the flexibility they allow, as the volume of water can be applied at every location at a given amount of time. Growers who have fixed irrigation equipment that are distributed throughout the irrigation area are in a much better situation to apply water precisely than growers with systems where the irrigated equipment is moved from one location to another. For example, drip and sprinkler irrigation systems, which have fixed equipment under the ground or over the field or permanent systems, have more flexibility in application than hand-moved sprinkler systems where farmers can only irrigate, say, 10% or 15% of the field at a time.

Sprinkler irrigation allows more flexibility than the traditional systems, but less than drip irrigation. However, the availability of equipment at the farm level is not the crucial determinant on the ability to have precise and flexible irrigation. Much depends on the allocation of water at the district level. When districts have a rigid canal system, a limited capacity to transfer water to different locations, and slow response time, a farmer will not gain much from having a flexible, modern irrigation system. In addition to the investment in modern irrigation systems, they also need to invest in storage systems that will enable them to adopt modern irrigation technologies.

The adoption of modern irrigation technologies, such as center-pivot and drip and sprinkler, were largely triggered by the high energy cost of pumping groundwater and water shortages during droughts. Reduction in water supplies during drought conditions, especially in California, forced water allocation systems to become more flexible. For example, the largest district in the state, the Westlands Water District, has installed a permanent set of volumetric pipes, making investments in modern technologies more profitable. When people started to adopt modern irrigation technologies, they gained more flexibility in terms of timing and location of applications and, once water districts introduced a more flexible system of water allocation, new improvements in irrigation were introduced. These improvements are much more consistent in spirit with precision farming.

Farmers complemented the adoption of modern irrigation technologies with management technologies that have two major components. First, they obtain climatic information above the levels of evaporation and other key variables that affect water utilization by crops. They rely on computer programs to optimize water allocation given this information. Climatic information has been obtained by several means. It may be obtained by equipment, including weather stations at the farm, or it may be obtained through hookups to a standardized weather station that belongs to the California Irrigation Management Information System (CIMIS) or a private network. The computer programs are used to manage reallocation of water within the farm system to determine how water flows within the farm, when and where, and how much to irrigate so water can be allocated optimally.

An alternative feature of some of the new irrigation systems is a type of monitoring equipment that is linked to a decision-making hardware or provides automated management instruction. For example, some farms use tensometers to detect soil moisture and environmental conditions. The information is transformed to an automated management system that can turn valves on and off to allocate water. In a conversation with Udi Sosnik, co-founder and marketing director of Orange Enterprises, Inc., a company that develops software for irrigation management in California, he suggests that there is significant heterogeneity in the use of automated irrigation programs and sophisticated irrigation schemes.

In some cases farmers have the personnel, knowledge, and resources to develop their own specialized software. These are mostly large sophisticated farms with the software to manage both irrigation and application of chemicals. In many cases the software is not only used for internal management of water within the farm but, most importantly, can be used to interact with the water districts to make decisions on how to obtain water from various sources.

Another group of growers will buy and utilize software and may make significant modifications using their own resources. A third group may not rely directly on software but, rather, follow advice from consultants. They may have contractual agreements or relatively frequent contacts with consultants who help guide their irrigation system. Some of the firms have a choice between hiring their own irrigation manager, who has the knowledge and experience to run irrigation on a farm, or to work with consultants who will provide advice for a fee.

Some farmers will obtain information from CIMIS and other sources and use a simple formula to adjust their irrigation systems. Studies on the adoption of CIMIS found that the use of CIMIS is significantly affected by the education of the farmer and the value of the crop. Educated farmers or growers, who have high-value crops, are more likely to adopt CIMIS and the most sophisticated irrigation technologies. Furthermore, in areas where water problems are more severe, adoption is more likely.

Adoption of the most sophisticated irrigation management system is likely to occur with high-value crops such as fruits and vegetables. Drip irrigation, for example, was introduced in the hills of southern California, and automated irrigation is more likely to occur in regions of southern

California where avocados and citrus, fresh tomatoes, and other vegetables are grown and, of course, in production of strawberries and flowers. A 2% or 3% increase in yield of these crops may make up for the extra investment in advanced irrigation. On the other hand, crops like cotton, with annual revenues of about $1,000 per acre are not likely be grown with the most sophisticated irrigation systems. Cotton growers, especially those in the west side of the Central Valley where water availability was constrained for long periods, are among the most enthusiastic and experienced users of irrigation systems and programs. They provide proof that even with cotton, a crop that does not generate high value, there is a great potential under the right circumstances to introduce technologies that substantially increase precision.

A better understanding of the factors that affect adoption of precision technology is achieved by a formal model of the decisions about variable input use and technology choices by farmers. Our initial assumption is that farmers are risk-neutral and aim to maximize their profits. They are aware of the variability in productivity across locations. This variability may be due to differences in fertility among locations as well as differences in input-holding capacity. For example, in some locations where the soil is sandy, water-holding capacity of fertilizers is very low. Traditional technologies entail equal uniform input applications at all locations, and modern technologies vary in applications to accommodate both fertility and input-holding capacity. Precision technologies will increase the variable profit, which is defined as revenue minus variable cost. However, for adoption to occur, this gain has to outweigh the extra fixed cost associated with modern technology. We show that property of the production function, as well as heterogeneity expressed by the distribution of both fertility and input-holding capacity and also the prices of output and input, will determine when farmers will adopt modern technology. Some parameters of production function, for example, the degree of responsiveness of output to changes in input at the various levels of production, will be quite crucial in determining the gain from adoption of precision technology and, thus, the likelihood of adoption.

A formal conceptual adoption model of precision farming is developed below. The model assumes certainty and profit maximization. We will also establish a foundation for an application of the model using a von Liebig production function. The result will be used in applications relying on data that are obtained from the Shafter Station in California. While conceptual analysis considers the case of one input, the more applied analysis will consider the case of fertilizers, pesticides, and water. The initial analysis will not consider environmental issues, but they will be introduced later.

4. MODELLING A PRECISION TECHNOLOGY

Assume a unit of land that is used to produce an agricultural crop with a variable input (say, water or chemical) and land quality is assumed to vary. There are two sources of variability and fertility, α, and input-holding capacity, β; α varies between $\underline{\alpha}$ to $\overline{\alpha}$ and values are from $\underline{\beta}$ to 1. A piece of land with $\alpha = \overline{\alpha}$, $\beta = 1$ is the most fertile and holds all the applied input. When β = .5, 50% of applied input is lost due to the runoff of percolation. Let x be input per unit of land and y be output per unit of land.

$$y = \alpha f(\beta x). \tag{1}$$

It is useful to define $e = \beta x$ as an effective input use, so let $y = f(e)$. The joint density functions of α and β is $g(\alpha, \beta)$ when

$$\int_{\underline{\alpha}}^{\overline{\alpha}} \int_{\underline{\beta}}^{1} g(\alpha, \beta)\, d\beta\, d\alpha = 1. \tag{2}$$

Assume that there are two technologies, and let i be the technology indicator, $i = 0$, for a traditional technology, and $i = 1$ for a modern precision technology. Each technology requires fixed costs. The annualized fixed cost of the precision technology is F_1, and it is assumed to be greater than the fixed cost of the traditional technology, F_0.

Each technology determines input use on each location, $x_i(\alpha, \beta)$. Total input under technology i is

$$X_i = \int_{\underline{\alpha}}^{\overline{\alpha}} \int_{\underline{\beta}}^{1} x_i(\alpha, \beta)\, g(\alpha, \beta)\, d\beta\, d\alpha, \quad \text{for } i = 0,1. \tag{3}$$

Output under technology i is

$$Y_i = \int_{\underline{\alpha}}^{\overline{\alpha}} \int_{\underline{\beta}}^{1} \alpha\, f(x_i(\alpha, \beta)\beta)\, g(\alpha, \beta)\, d\alpha\, d\beta, \quad i = 0,1. \tag{4}$$

Profit under technology i is

$$\Pi_i = pY_i - wX_i - F_i,$$

where p and w are output and input prices, respectively.

Under the modern precision technology, the variable input is adjusted to changes in conditions across locations, reflected by changes of the quality parameters α and β. The input at each location is denoted by the function $x_1(\alpha, \beta)$, and the profit-maximization problem with the precision technology is

$$\int_{\underline{\alpha}}^{\bar{\alpha}}\int_{\underline{\beta}}^{1} \underset{x_1(\alpha,\beta)}{Max}\left\{p\alpha\ f\left(x_1(\alpha,\beta)\beta\right) - wx_1(\alpha,\beta)\right\}\ g(\alpha,\beta)\,d\alpha\,d\beta - F_1. \tag{5}$$

subject to: $x_1(\alpha,\beta)\geq 0$ and $p\alpha\ f\left(x_1(\alpha,\beta)\beta\right) - wx_1(\alpha,\beta)\geq 0.$

The choice of input use for any location is subject to a non-negativity constraint on both the input use and the per acre profits. When the optimal input use is positive, it occurs where the value of marginal product of the input is equal to its price,

$$p\beta\alpha f'\left(\beta x_1(\alpha,\beta)\right) = w, \tag{6}$$

where $f'(e_1)$ is the marginal productivity of effective input, $e_1 = \beta x_1(\alpha,\beta))$. The comparative statics of Caswell and Zilberman (1985) is useful to predict patterns of changes of output and applied per unit of land in response to changes of β and α. Land with higher soil fertility (α) will tend to have higher input use and output per acre. Higher input-holding capacity (β) leads to increased yields. Higher β will save input if the production function is very convex (the elasticity of marginal productivity (EMP $= -f''(e_1)e_1 / f'(e_1)$ >1). If the EMP < 1, the marginal productivity of the applied input declines slowly, and increase in input use efficiency will actually increase both input use and output. The results of Caswell and Zilberman also suggest that subsections of the field with either low fertility and/or input-holding capacity will not be utilized. The results also confirm that supply with the precision technology and the demand for the variable input will increase as output price increases and input price declines.

While precision technology allows variation of inputs in response to changes in land conditions, the variable input level is applied uniformly with the traditional technology, and applied input per unit of land is denoted by x_0. It is determined solving

$$\Pi_0 = \underset{x_0}{Max}\int_{\underline{\alpha}}^{\bar{\alpha}}\int_{\underline{\beta}}^{1}\left\{p\alpha\ f(x_0\beta) - wx\right\}g(\alpha,\beta)\,d\alpha\,d\beta\ -\ F_0 \tag{7}$$

subject to $\Pi_0 \geq 0,\ x_0 \geq 0$.

The optimal level of applied input, when it is positive, is determined solving

$$p\int_{\underline{\alpha}}^{\bar{\alpha}}\alpha\int_{\underline{\beta}}^{1}\beta f'(x_0\beta)\,g(\alpha,\beta)\,d\beta\,d\alpha = w. \tag{8}$$

The left-hand side of (8) is the average value of marginal product of the input over all qualities, and the optimal x_0 is determined so it is equal to the variable input price w. Input use and output will increase in response to an increase in p or a decline of w. With the traditional technology, input will be applied on subfields where it is generating losses, while in the productive section of the fields, profit opportunities will be lost because of underapplication of the input.

Using the simplifying assumption that $F_1 - F_0$ represents the difference in the annualized fixed cost resulting from the adoption of precision farming, a profit-maximizing farmer will adopt the technology if

$$p[Y_1 - Y_0] - w[X_1 - X_0] - F_1 - F_0 > 0.$$

It is likely that the adoption of the precision farming will increase the fixed cost ($F_1 - F_0 > 0$). Thus, adoption of precision farming will be because of its impact on operational profits $p[Y_1 - Y_0] - w[X_1 - X_0]$. In some cases precision farming will both increase yield and reduce costs. However, it actually may reduce output and yet be profitable because of the cost-saving effect. Adoption is likely to be advanced as the cost of the precision technology declines. Cohen-Vogel and Zilberman (2001) showed that adoption is also likely to increase as input price increases when the cost-saving effect is significant. If the technology has a yield-increasing effect, adoption is likely to increase at periods of high demand and output prices. But when precision technology actually reduces output, its likelihood of adoption may increase when output price declines, especially when it is also associated with input price increase.

The actual impacts of precision technologies depend on the specification of the crop production function, the prices of the output, the variable input and the fixed cost of the technology, and the distribution of the sources of heterogeneity within a field. As the analysis of Cohen-Vogel and Zilberman (2001) demonstrates, it is difficult to generalize both the direction and magnitudes of impacts of adopting precision farming. Therefore, we assess below the impacts of switching to precision technology in a specific case, cotton production in California, and assume a von Liebig production function. Much of the effort to introduce precision technology has been in cotton and, while parameters vary across regions, the California figures are likely to provide good indicators on the value of precision technology elsewhere.

5. THE VON LIEBIG CASE

Let us first consider a special case, a von Liebig production function. In this case the production function has a fixed proportion for a given range of effective input, and it reaches maximum yield for $e = \bar{e}$. For a given α and β,

$$y = \begin{cases} \alpha e \text{ for } 0 \le e \le \bar{e} \\ \alpha\bar{e} \text{ for } e \ge \bar{e}. \end{cases} \tag{9}$$

This production function is depicted in Figure 1 for $\alpha = 1$.

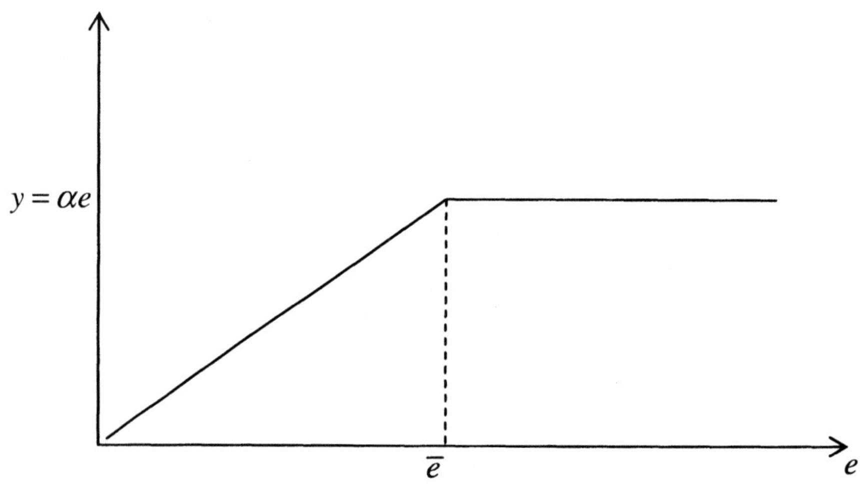

Figure 1. Von Liebig production function

In the case of cotton in California, maximum yield per acre is between 1,000 and 1,600 pounds with effective water of 2.6 AF (acre-foot) per acre (Hanemann et al., 1987). The minimum level of effective water required is 1.3 AF per acre. Thus, in cases when maximum yield is 1,300, $\alpha = 500$ and the production function is

$$y = \begin{cases} 500e \quad \text{for } 0 \le e \le 2.6 \\ 1,300 \quad \text{for} \quad e \ge 2.6 \end{cases}.$$

The value of α may vary from 400 (in which case maximum output is 1,040 pounds per acre) to 650 (yielding maximum yield of 1,670 pounds per acre).

Now irrigation efficiency assumes values from .3 to .4. Thus, when $\beta = .4$, the maximum level of applied water is 6.5 (2.6/.4), while with $\beta = .9$, it is approximately 2.9 per acre. When the same level of water is applied at all locations, yield monitoring can detect the parameter at much of the field. The detection, however, is far from perfect, since yield monitoring detects on parameter, while there are two sources of variability.

To illustrate this point, suppose 3.9 AF per acre is applied throughout the field. All the lands with $\beta \geq 2/3$ will yield their maximum yield, which is 1,300 pounds per acre and, for locations with $.4 < \beta < 2/3$, output per acre is $3.9 * 500 * \beta$.

Let's first find $x_1(\alpha,\beta)$, input use per unit of land under precision technology, for the Von Liebig production function per acre f presented in (9). Since fixed cost is assigned to the whole field, not every unit of land profit per unit of land $\pi_1(\alpha,\beta)$ is

$$\pi_1(\alpha,\beta) = \begin{cases} p\alpha x_1 \cdot \beta - w x_1 & \text{if } 0 \leq x_1 \leq \dfrac{\bar{e}}{\beta} \\ p\alpha\bar{e} - w x_1 & \text{if } x_1 \leq \dfrac{\bar{e}}{\beta} \end{cases} . \tag{10}$$

The nature of the technology suggests that, when input is applied, maximum yield is attained with $x = \bar{e}/\beta$. However, in some cases revenues may not cover input cost so that

$$x_1(\alpha, \beta) = \begin{cases} 0 & \text{if } p\,\alpha\bar{e} - w\bar{e}/\beta < 0 \\ \bar{e}/\beta & \text{if } p\,\alpha\bar{e} - w\bar{e}/\beta \geq 0 \end{cases} . \tag{11}$$

Thus, no input will be applied when $\beta < \dfrac{w}{p\alpha}$, and there will be a range of land qualities that will not be utilized with precision farming. For every level of input-holding capacity, β, there is a minimum $\alpha, \alpha_L(\beta)$,

$$\alpha_L(\beta) = \frac{w}{\beta p}, \tag{12}$$

so that the location with the same β but with $\alpha > \alpha_L(\beta)$ will operate, and the location with $\alpha < \alpha_L(\beta)$ will not be used. Equation (12) suggests that, where the variable input price is sufficiently high, $w > p\alpha\beta$, the land will not be used.

Condition (10) is used to determine the output, input use, and profit under precision technology. When $\beta = 1$, only locations with $\alpha > \alpha_L(1) = w/p$ will operate, and $\alpha_L(1)$ is the lowest that will be utilized. For a given α, the lowest β with positive production is $\beta_L(\alpha) = w/p\alpha$, and production occurs for $\beta_L(\alpha) \leq \beta \leq 1$.

Using these definitions, assuming that $\underline{\alpha}$ is sufficiently high,

$$X_1 = \int_{\underline{\alpha}}^{\bar{\alpha}} \int_{Max\{\beta_L(\alpha),\underline{\beta}\}}^{1} \frac{\bar{e}}{\beta} g(\alpha,\beta)\, d\beta d\alpha \tag{13}$$

$$Y_1 = \int_{\underline{\alpha}}^{\bar{\alpha}} \int_{Max\{\beta_L(\alpha),\underline{\beta}\}}^{1} \alpha\bar{e}\; g(\alpha,\beta) d\beta d\alpha \tag{14}$$

$$\pi_1 = \int_{\underline{\alpha}}^{\bar{\alpha}} \int_{Max\{\beta_L(\alpha),\underline{\beta}\}}^{1} \left[p\alpha\bar{e} - \frac{w\bar{e}}{\beta} \right] g(\alpha,\beta) d\beta d\alpha - F_1 . \tag{15}$$

Equations (13)–(15) demonstrate that increase in the upper bound of effective input per unit of land ($\bar{e}$) tends to increase water use, aggregate output, and profits, both through a direct effect (increase values per each unit of land) and, indirectly, through increasing the share of the field to be utilized (reducing $\beta_L(\alpha)$). Changes in p and w have the expected effects ($dX_1 / dw \le 0, dY_1 / dw \le 0, d\pi_1 / dw \le 0, dX_1 / dp \ge 0, dY_1 / dp \ge 0,$ $d\pi_1 / dp \ge 0$).

The derivation of outcomes without precision farming requires a uniform amount of water x_0 to be applied throughout the field. For locations with $\alpha > e / x_0$, the marginal contribution of water is zero. The output under uniform application is

$$Y_0 = \int_{\underline{\alpha}}^{\bar{\alpha}} \int_{\underline{\beta}}^{\bar{e}/x_0} \alpha\beta x_0 g(\alpha,\beta)\, d\beta d\alpha + \int_{\underline{\alpha}}^{\bar{\alpha}} \int_{\bar{e}/x_0}^{1} \alpha\bar{e} g(\alpha,\beta)\, d\beta d\alpha . \tag{16}$$

Thus, the profits with the traditional technology is

$$\pi_0 = pY_0 - wx_0 - F_0 . \tag{17}$$

Maximization of (17) yields the first-order condition

$$p \int_{\underline{\alpha}}^{\bar{\alpha}} \int_{\underline{\beta}}^{\bar{e}/x_0} \alpha\beta\; g(\alpha,\beta)\; d\alpha d\beta = w . \tag{18}$$

A marginal increase in x_0 will increase output by $\alpha\beta$ unit and revenue by $p\alpha\beta$ in all locations when input use is not at the limit $\left(x_0\beta < \bar{e}\right)$, and at the optimal level of applied water this marginal gain has to be equal to the price of the input w.

With uniform water use, these are locations with high α and β, which are over-irrigated $\left(x_0\beta > \bar{e} \text{ or } p\alpha\beta x_0 < w\right)$, and then locations with low α and β, which are under-irrigated $\left(x_0\beta < \bar{e} \text{ and } p\alpha\beta x_0 \ge w\right)$.

Assume that α and β have two independent uniform distributions. The means of α and β are $E(\alpha) = (\bar{\alpha} + \underline{\alpha})/2$ and $(1+\underline{\beta})/2$, respectively, and their joint density is

$$g(\alpha,\beta) = \frac{1}{(\bar{\alpha}-\underline{\alpha})(1-\underline{\beta})}$$

when $\underline{\beta}$ is the minimum value of β. The first-order condition (18) becomes

$$p\int_{\underline{\alpha}}^{\bar{\alpha}}\int_{\underline{\beta}}^{\bar{e}/x_0} \alpha\beta \frac{1}{(\bar{\alpha}-\underline{\alpha})(1-\underline{\beta})} d\beta d\alpha = w.$$

This condition becomes

$$\frac{p}{4}\frac{(\bar{\alpha}+\underline{\alpha})}{(1-\underline{\beta})}\left(\frac{\bar{e}^2}{x_0^2} - \underline{\beta}^2\right) = w,$$

which yields

$$X_0 = x_0 = \frac{\bar{e}}{\sqrt{\underline{\beta}^2 + 4\frac{w}{p}\frac{(1-\underline{\beta})}{(\underline{\alpha}+\bar{\alpha})}}}. \quad (19)$$

Equation (19) suggests that the water application, when the price of water is zero, is $\bar{e}/\beta$; and in this case we reach maximum output, which is $\bar{e}(\bar{\alpha}+\underline{\alpha})/2$. The denominator of equation (19) suggests that when water price is positive, $x_0 < \bar{e}/\underline{\beta}$ and some part of the field will produce less than maximum yield $(\alpha\bar{e})$. The difference in $\bar{e}/\beta - x_0$ increases as the price of water increases and decreases so output price and average productivity $(\bar{\alpha}+\underline{\alpha})/2$ increase. The output under the traditional technology is

$$Y_0 = \int_{\underline{\beta}}^{\bar{e}/x_0} \frac{\bar{\alpha}+\underline{\alpha}}{2}\frac{\beta x_0}{(1-\underline{\beta})}d\beta + \int_{\bar{e}/x_0}^{1}\frac{\bar{\alpha}+\underline{\alpha}}{2}\frac{\bar{e}}{(1-\underline{\beta})}d\beta$$

$$= E(\alpha)\left\{x_0\frac{(\bar{e}/x_0+\underline{\beta})}{2}\frac{(\bar{e}/x_0-\underline{\beta})}{1-\underline{\beta}} + \bar{e}\frac{(1-\bar{e}/x_0)}{1-\underline{\beta}}\right\}. \quad (20)$$

Total output is a weighted sum of the output produced by lands with low β's (the first item in the right-hand side of (20)) and lands with higher water-holding capacity, which reach the land productivity constraint.

Equations (19) and (20) suggest that output with the traditional technology will increase as output price, expected land fertility, and the upper bound on effective water ($\bar{e}$) increase and water price declines. The output of the traditional technology is smaller than of the precision technology when all the land is utilized. Solving (14) for the case of uniform distribution when $\underline{\beta} \geq \beta_L(\alpha) = w / p\alpha$ results in output level $Y_1 = E(\alpha)\bar{e} > Y_0$. The water use in this case is solved from equation (13) to be $X_1 = -\bar{e}\ \ln(\underline{\beta})/(1-\underline{\beta})$.

Table 1 presents the results of a simulation that attempts to estimate the potential gain of precision farming in irrigation. It considers cases where $350 < \alpha < 700$, $\underline{\beta} < \beta < 1$, and several values of Table 1 present the results of a simulation that attempts to estimate the potential gain of precision farming in irrigation. It assumes uniform distributions of both α and β. In all the simulations, α is distributed between 350 and 700. In some of the simulations, β is distributed between .4 and 1; and in others, the variability of β is smaller. We assume three scenarios of maximum effective water - $\bar{e}$ can assume values of 2.6, 2.7, and 3.0 AF per acre. With these parameters, maximum yield per acre occurs where $\bar{e}$ is 3 and α is 700, and yield per acre is 2,100 lbs./acre. The lowest yield per acre is 910 lbs./acre, corresponding to α = 350 and $\bar{e}$ = 2.6. We also consider two output prices, .70 cents and .90 cents/lb., and water prices of $40 and $60/AF.

The results suggest that the adoption of precision farming saves water and increases yield. The water-saving effect is higher when the water price is actually low and output price is high. Generally the water-saving effect is between 10% and 20%. There is also a yield effect that can reach up to 33%, and the yield effect is high when water price is low and output price is high.

The overall gain from adoption increases with output price. The strong effect of an output price increase is apparent when comparing rows 1 and 2. Similarly, a comparison of rows 1 and 3 or rows 2 and 4 demonstrates that a higher input price leads to increase in gain from adoption, which is obvious when gains from adoption of precision also increase where the variability increases. That is clear from a comparison of row 1, which has more variability, with rows 13 and 25. As variability declines, the gain from adoption declines. The highest potential gain from adoption is row 12, which corresponds to the highest effective water, water price, output price, and variability of β. The results show how the potential gain from adoption is sensitive to changes in key parameters of the system.

Table 1: The potential of precision farming in the irrigation of cotton

	Parameters						**Basis**			**With precision**			
Row	Maximum effective water	Water price	Low β	Output price	Low α	High α	Water	Yield	Profit	Water	Yield	Profit	Delta profit
1	2.6	40	0.4	0.7	350	700	4.82	1186	638	3.97	1365.0	797	159
2	2.6	40	0.4	0.9	350	700	5.08	1224	899	3.97	1365.0	1213	314
3	2.6	60	0.4	0.7	350	700	4.36	1105	512	3.97	1365.0	940	428
4	2.6	60	0.4	0.9	350	700	4.65	1159	764	3.97	1365.0	1213	449
5	2.7	40	0.4	0.7	350	700	5.01	1232	662	4.12	1417.5	975	313
6	2.7	40	0.4	0.9	350	700	5.28	1271	933	4.12	1417.5	1259	326
7	2.7	60	0.4	0.7	350	700	4.53	1147	531	4.12	1417.5	975	444
8	2.7	60	0.4	0.9	350	700	4.83	1203	793	4.12	1417.5	1259	466
9	3.0	40	0.4	0.7	350	700	5.56	1369	736	4.58	1575.0	1082	346
10	3.0	40	0.4	0.9	350	700	5.87	1413	1037	4.58	1575.0	1397	360
11	3.0	60	0.4	0.7	350	700	5.03	1274	590	4.58	1575.0	1082	491
12	3.0	60	0.4	0.9	350	700	5.37	1337	881	4.58	1575.0	1397	515
13	2.6	40	0.6	0.7	350	700	3.89	1242	714	3.32	1365.0	823	109
14	2.6	40	0.6	0.9	350	700	3.98	1269	983	3.32	1365.0	1217	234
15	2.6	60	0.6	0.7	350	700	3.71	1181	604	3.32	1365.0	944	340
16	2.6	60	0.6	0.9	350	700	3.83	1222	870	3.32	1365.0	1217	348
17	2.7	40	0.6	0.7	350	700	4.04	1290	741	3.45	1417.5	980	239
18	2.7	40	0.6	0.9	350	700	4.13	1318	1021	3.45	1417.5	1264	243
19	2.7	60	0.6	0.7	350	700	3.85	1227	627	3.45	1417.5	980	353
20	2.7	60	0.6	0.9	350	700	3.97	1269	903	3.45	1417.5	1264	361
21	3.0	40	0.6	0.7	350	700	4.49	1433	824	3.83	1575.0	1088	264
22	3.0	40	0.6	0.9	350	700	4.59	1464	1134	3.83	1575.0	1403	268
23	3.0	60	0.6	0.7	350	700	4.28	1363	697	3.83	1575.0	1088	391
24	3.0	60	0.6	0.9	350	700	4.42	1410	1004	3.83	1575.0	1403	399
25	2.6	40	0.8	0.7	350	700	3.14	1272	765	2.90	1365.0	839	75
26	2.6	40	0.8	0.9	350	700	3.17	1293	1037	2.90	1365.0	1220	183
27	2.6	60	0.8	0.7	350	700	3.10	1226	672	2.90	1365.0	947	275
28	2.6	60	0.8	0.9	350	700	3.13	1257	943	2.90	1365.0	1220	277
29	2.7	40	0.8	0.7	350	700	3.27	1321	794	3.01	1417.5	983	189
30	2.7	40	0.8	0.9	350	700	3.29	1343	1077	3.01	1417.5	1267	190
31	2.7	60	0.8	0.7	350	700	3.21	1273	698	3.01	1417.5	983	285
32	2.7	60	0.8	0.9	350	700	3.25	1305	980	3.01	1417.5	1267	287
33	3.0	40	0.8	0.7	350	700	3.63	1468	882	3.35	1575.0	1091	209
34	3.0	40	0.8	0.9	350	700	3.65	1492	1196	3.35	1575.0	1406	210
35	3.0	60	0.8	0.7	350	700	3.57	1414	776	3.35	1575.0	1091	315
36	3.0	60	0.8	0.9	350	700	3.61	1450	1089	3.35	1575.0	1406	318

6. THE SIMULATION RESULTS IN PERSPECTIVE

The simulation results illustrate the potential of precision technology to increase both water productivity and agricultural productivity. The uniform distribution was used for its simplicity. However, it may lead to overstatement of the gains from precision because it is the distribution with high degree of variability (compared to unimodal distributions), but other aspects of the simulation underemphasized other sources of gains from precision.

First, we only considered the impacts of precision on water use. Currently, precision technologies are mostly used in California to save on fertilizers as well as improve the use of growth regulators, to achieve uniformity of yield, and to assure that maximum potential of output is obtained at harvest time. The same modeling principles that we used to assess the impact of precision farming in irrigation will be used to assess the impacts of using precision to apply fertilizers and growth regulators. Appropriate use of growth regulators in locations within fields where growth has to be slowed or enhanced can increase yield by 20% or 30% in some areas, and the capacity to identify these locations is crucial.[1] Uniform applications of nitrates may result in waste in some locations and underuse in others, and the overall relative order of magnitudes of impacts that we saw in water applies to other inputs.

Second, in the future it is likely that new development in biotechnology will contribute to the effectiveness of precision farming. With genetic modification, it will be possible to produce varieties that are slightly different and that will adjust to field conditions, so that combining the expansion of varietal choice with equipment that can allocate different varieties to different locations will provide a new source of expanded precision. Our analysis suggests that this expansion has the potential to contribute to higher output as well as input savings.

Third, our analysis ignores the potential environmental gain from precision farming. As Khanna and Zilberman (1997) argue, residue input can be a source of pollution. If pollution is priced, one of the advantages of precision farming will be to reduce pollution levels, for example, by reducing input use in locations with low water-holding capacity (β's). Note that the simulation considered cases that, even with the lowest β's, production is worthwhile. One can envision cases where in some locations the water-holding capacity is so low that some land is taken out of production. A taxation of pollution will increase this range of subfields that are not utilized with precision farming, and that will reduce the yield effect of precision farming but increase the water-saving effect.

The paper presents a conceptual framework that has to be refined and tailored to specific situations to be applicable. Some of the areas where the analysis has to be expanded include:

(1) Development of empirical distributions of different sources of variability within fields to assess how the impacts of precision will vary under different distributional assumptions.

(2) Analysis of the implication of finance on adoption of precision farming. As Khanna et al. (1999) have shown, the annualized cost of precision farming varies depending on whether you rent or purchase it.

(3) Consideration of reliability issues. When precision equipment makes diagnostic errors, its effectiveness declines. It is important to analyze the profitability of decision systems given reasonable estimates of misdiagnosis.

(4) Expansion to consider learning and dynamics. The analysis here is static. In reality, adoption of precision farming entails significant learning by doing and adjustment of the behavior in response to key parameters as experience is accumulated. The initial gains from precision farming will in most cases be below the long-term expected gain. Our simulation provides an estimate of upper bound and long-run gain, and one may recognize that it will take several years to reach this gain. A more realistic estimation of the benefits of precision farming will factor in a learning-by-using process and its implications.

The analysis of precision farming should also consider alternative functional forms besides von Liebig. It will be useful to develop field-level production function with and without precision technology when the microlevel production function is quadratic, Cobb-Douglas, etc. Different combinations of microlevel production functions and distribution of sources of heterogeneity can be combined using techniques introduced by Felipe and Fisher (2003) yield interesting field-level production functions with and without precision. Furthermore, the analysis can be expanded by incorporating random factors that are sources of risk. In these cases, the riskiness of the production function may lead to adoption models that explicitly consider risk aversion.

Studies on adoption of precision farming that affect water use should consider the issues of water right ownership and capacity to trade and also investigate the impact of input and output prices. Sometimes the key to adoption of precision farming is removal of barriers to trade and water rights and construction of capacity that facilitates water transfers. Institutional factors that include water rights regimes, availability of credit, and support for new technologies should be considered in assessing the likelihood and potential of adoption of precision irrigation in real-life situations.

Notes

1. According to Bruce Roberts of Fresno State University.

References

Burt, O. R., 1964. 'The Economics of Conjunctive Use of Ground and Surface Water', *Hilgardia*, 36(2), 31-111.

Caswell, M. and D. Zilberman, 1985. 'The Choices of Irrigation Technologies in California', *American Journal of Agricultural Economics,* 67(2), 224-239.

Cohen-Vogel, D.R. and D. Zilberman, 2001. *'Response to Intra-Firm Variability: Production and Adoption Impacts of a Precision Technology', Working Paper, Department of Agricultural and Resource Economics, University of California, Berkeley, California.*

Feinerman, E., J. Letey and H. J. Vaux, Jr., 1983. 'The Economics of Irrigation with Non-Uniform Infiltration', *Water Resources Research,* 19, 1410-1414.

Felipe, J. and F.M. Fisher, 2003. 'Aggregation in Production Functions: What Applied Economists Should Know', *Metroeconomica,* 54(2-3), 208-262.

Gardner, B.D., 1995. 'Irrigation Subsidies: Some Efficiency, Equity, and Environmental Implications', *The Economics of Agriculture*. Volume 2: *Papers in Honor of D. Gale Johnson,* edited by J.M. Antle and D.A. Sumner, University of Chicago Press, Chicago and London, pp. 300-319.

Grote, K., S. Hubbard and Y. Rubin, 2003. 'Field-Scale Estimation of Volumetric Water Content Using GPR Groundwave Techniques', *Water Resources Research,* 39(11), SBH5-1—SBH5-11.

Hanemann, M., E. Lichtenberg, D. Zilberman, D. Chapman, L. Dixon, G. Ellis and J. Hukkinen, 1987. 'Economic Implications of Regulating Agricultural Drainage to the San Joaquin River', *Regulation of Agricultural Drainage to the San Joaquin River.* SWRCB Order No. W.G. 85-1. Technical Committee Report. Appendix G (two vols.). Sacramento, California: State Water Resources Control Board.

Khanna, M., O.F. Epouhe, and R. Hornbaker, 1999. 'Site Specific Crop Management: Adoption Patterns and Incentives', *Review of Agricultural Economics*, 21(2), 455-472.

Khanna, M. and D. Zilberman, 1997. 'Incentives, Precision Technology and Environmental Quality', *Ecological Economics,* 23(1), 25-43.

National Research Council. Committee on Assessing Crop Yield: Site-Specific Farming, Information Systems, and Research Opportunities, 1997. *Precision Agriculture in the 21st Century: Geospatial and Information Technologies in Crop Management,* Washington, D. C.: The National Academies Press.

Zilberman, D., N. MacDougall and F. Shah, 1994. 'Changes in Water Allocation Mechanisms for California Agriculture', *Contemporary Economic Policy*, 12(1), 122-133.